THE INCREDIBLE JOURNEY OF PLANTS

· *The* ·

INCREDIBLE JOURNEY
of PLANTS

Stefano Mancuso

Watercolors by Grisha Fischer
Translated from the Italian by Gregory Conti

OTHER PRESS
NEW YORK

First softcover edition 2022
ISBN 978-1-63542-191-0

Originally published in Italian as *L'incredibile viaggio delle piante*
in 2018 by Editori Laterza, Rome.
Copyright © 2018 Gius. Laterza & Figli
English translation copyright © 2020 Gregory Conti

Production editor: Yvonne E. Cárdenas
Text designer: Jennifer Daddio / Bookmark Design & Media Inc.
This book was set in Horley OS MT

1 3 5 7 9 10 8 6 4 2

LIBRARY OF CONGRESS CATALOGING-IN-PUBLICATION DATA

Names: Mancuso, Stefano, author. | Conti, Gregory, 1952- translator.
Title: The incredible journey of plants / Stefano Mancuso ; watercolors by
Grisha Fischer ; translated from the Italian by Gregory Conti.
Description: New York : Other Press, 2020. |
Includes bibliographical references.
Identifiers: LCCN 2019027763 (print) | LCCN 2019027764 (ebook) |
ISBN 9781635429916 (hardcover) | ISBN 9781635429923 (ebook)
Subjects: LCSH: Plants–Migration.
Classification: LCC QK101 (print) | LCC QK101 (ebook) | DDC 581.3–dc23
LC record available at https://lccn.loc.gov/2019027763
LC ebook record available at https://lccn.loc.gov/2019027764

TO ROSARIA AND FRANCO

my parents

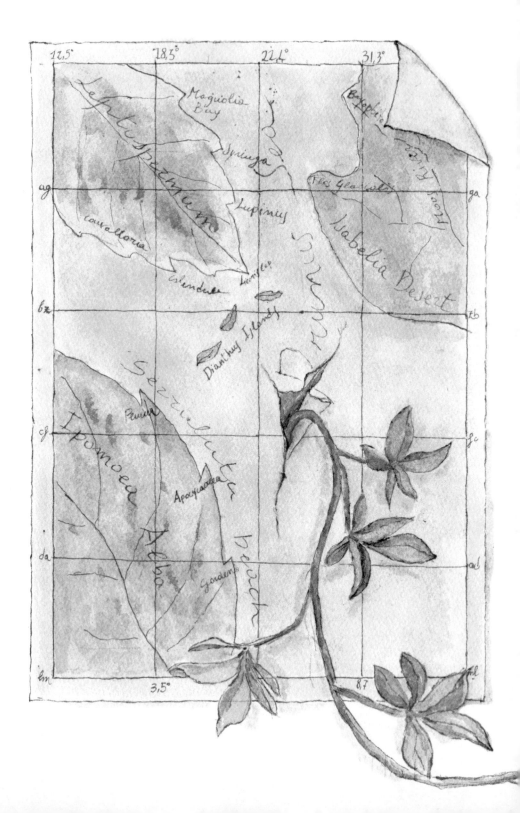

· CONTENTS ·

· PREFACE ·

Do you remember the Frank Capra masterpiece *It's a Wonderful Life,* with James Stewart as the unforgettable George Bailey? I really hope you all know it. The plot of the film is very simple: it's centered around the sacrifices of dreams and aspirations that the hero, George Bailey, makes his whole life long just for the sake of helping others.

As a child, George saves his brother, Harry, from drowning in a frozen pond, becoming deaf in one ear in the process. As an adult, he gives up his own dreams in order to manage the small building and loan founded by his father. George gives his college tuition money to brother Harry so that Harry can go to college. When George gets married, in 1929, the year of the Wall Street crash, he uses the money he had saved for his honeymoon to keep the building and loan solvent and avoid bankruptcy. Sacrifice after sacrifice, George's life goes by unobserved until, owing to a series of events I won't go into, our hero decides to kill himself. He is about to throw himself into the river when Clarence (Angel Second Class) saves him, transports him into a parallel world, and shows George what the real world would have been like had he never been born.

I know: when it's recounted like this it's tough to keep a straight face, but Capra is able to take an edifying Christmas story and turn it into one of the milestones of cinema history. Actually, now that I've talked to you about it, I can't wait for Christmas so I can have the chance to see it again.

Well, plants are the George Baileys of this planet. Nobody pays them respect; they are studied much less than animals; we have no more than a vague idea of how many of them there are, how they work, what their characteristics are. Yet, without them, the life of each of us animals would not be possible. It would be instructive if a master of cinema of the caliber of Frank Capra could show us one day what our world would be like if plants had never been born.

We know very little about plants, and, quite often, the little we think we know is wrong. We are convinced that plants are not able to perceive the environment around them, while in reality, quite to the contrary, they are more sensitive than animals. We are sure that plants belong to a silent world, deprived of the ability to communicate, but, instead, plants are great communicators. We are convinced that they don't carry on any kind of social relationship, but, quite the opposite, they are exquisitely social organisms. We are, above all, absolutely certain that plants are immobile. On this point, we are immovable. Plants do not move; after all, just look at them. Isn't the big difference between animal (that is, animated, endowed with movement) and vegetable organisms exactly that?

Actually, we're wrong about this one, too. Plants are not immobile at all. They move a lot, only at a slower pace. Plants are not unable to move. What plants are unable to do is *locomote*, or move from place to place, at least not in the course of their lifetimes. The adjective that defines them, in fact, should not be "immobile" but "sessile," or, if you prefer, "rooted." A sessile organism cannot move from the place where it was born, but it can move, how and as much as it likes. That's what plants do, and each of us can understand that just by taking a look at one of the thousands of time-lapse videos that can be found all over the internet.

Although plants are not able to change places over the course of their individual lives, they are able, from generation to generation, to conquer the most distant lands, the most impervious areas, and the regions least hospitable to life, with a tenacity and capacity for adaptation that have often left me envious.

As I have written elsewhere, plants are incredibly different from animals. Their bodies, their architecture, their strategies are often diametrically opposed to those of animals. Animals have one command center; plants are multicentric. Animals have single or double organs; plants have diffuse organs. Animals are individuals (in the sense of being indivisible); plants are not individuals at all, being more like colonies. In a word, it would seem that the emphasis in animals is more on the singular, while in plants it's more on the plural. In animals, the individual counts more; in plants, the group. Organisms so different from us must be observed through the lens of comprehension or inclusion, not through the lens of similitude. We will never be able to understand plants if we look at them as if they were impaired animals. They are a form of life that is *different*, neither simpler nor less developed, than the animal form of life.

If they are looked at through eyes without an animal filter, their extraordinary characteristics emerge clearly and indubitably, everywhere, even in areas where it would seem unlikely, as in their ability to move. When it comes to migrants and migration, it is helpful to study plants in order to understand that we are talking about an unstoppable phenomenon. Generation after generation, using spores, seeds, or any other means, vegetables move about and advance through the world, conquering new spaces. Ferns release astronomical quantities of spores that can be transported by the wind for thousands of miles, for years and years. The number and variety of instruments by way of which seeds are dispersed in the environment are astounding. It is as if, over the course of evolution, every single possibility has been taken into consideration and, at one time or another, each of the tested solutions has found some species ready to try it.

Thus, we have seeds dispersed by wind; by turning the soil; dispersed by animals in general or by specific groups of animals, such as ants, birds, or mammals; dispersed by animals through ingestion or by sticking to their fur; dispersed by water or simply by dropping from the plant; dispersed by the undulating movement of the mother plant or thanks to propulsive mechanisms; by desiccation of fruit or, on the contrary, by hydration; and who knows how many other ways I've left out. Every year brings discoveries of different sophisticated strategies developed by plants

to improve their seeds' chances of germination. In the variety of
ways, procedures, and means, we can glimpse life's unceasing
impulse to expand, the impulse that has driven plants to colonize
every possible environment on Earth.

The story of this unstoppable expansion is largely unknown:
how plants convince animals to transport them around the
world; how some plants need the help of specific animals in
order to spread; how plants have been able to grow in places
so inaccessible and inhospitable as to be left on their own and
isolated; how they have held up against the atomic bomb and the
disaster at Chernobyl; how they are able to bring life to sterile
islands; how they manage to travel through the ages; how they
navigate around the world. These are just some of the stories told
in the pages that follow. Stories of pioneers, fugitives, veterans,
combatants, hermits, and lords of time await us. I won't prolong
the wait.

Rhamnus
Acanthus
Liliaceae
Adonis Ocean
Actinidia Ocean
Allium

...ero

Vitex

Julibrijjin

Agnus

coltus

Azalea

Pyzacantha

coccinea

Atonis
Ocean

pennica

Agezatum

Oxalis Sea

...eum

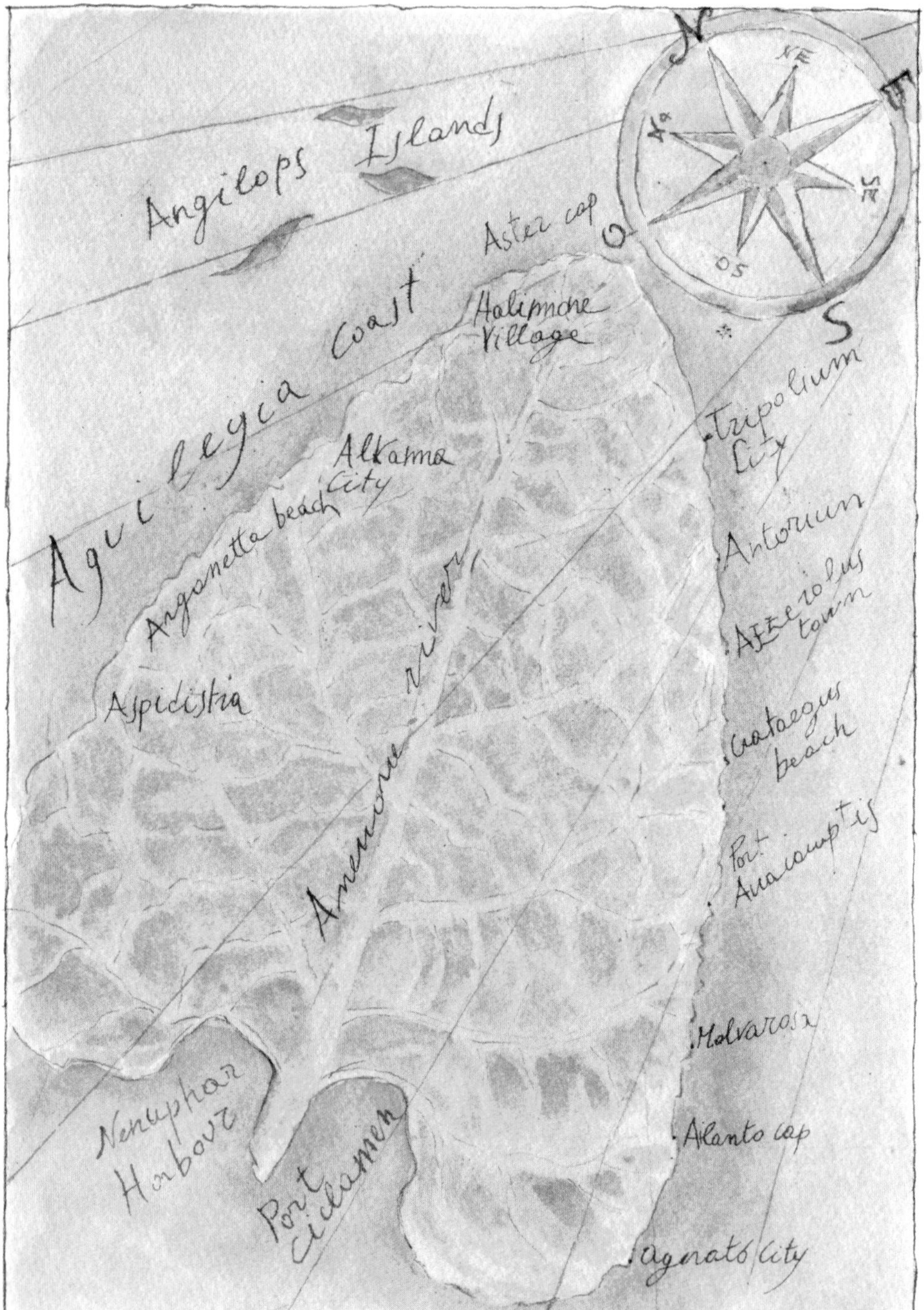
N
S
O
Angilops Islands
Aster cap
Aquilegia Coast
Halimione Village
Alkanna City
Aspidistra
Anemone river
Tripolium City
Crataegus beach
Port Anacamptis
Malvarosa
Ailanto cap
Agerato City
Nenuphar Harbour
Port Ciclamen

. 1 .

PIONEERS, COMBATANTS, AND VETERANS

---

type species Weeping Willow – domain Eukaryota – kingdom Plantae

division Magnoliophyta – class Magnoliophyta – order Salicales

family Salicaceae – genus *Salix* – species *Salix babilonica*

origin China – diffusion Worldwide

first appearance in Europe Seventeenth century

For me, the word "pioneer" evokes the epic of the West and the adventure stories of the American frontier. I think that's true for a lot of people. Someone says "pioneer" and a switch turns on in my memory, lighting up the faces of Gregory Peck, John Wayne, James Stewart, Eli Wallach, Richard Widmark, Lee Van Cleef, Henry Fonda, Debbie Reynolds, and, of course, Karl Malden, with his big broken bulbous nose, from the incredible cast of *How the West Was Won*. For me, "pioneer" means Salgari novels and cowboy movies, and that's it. To others, though not many, it might recall the special units of armies that, since ancient times, have been involved in opening up new roads for the arrival of the

1

infantry. But very few, and maybe nobody, will associate the word "pioneer" with plants.

This is really unfair. Plants should be the first thing to come to mind when we talk about pioneers, not the stars of Hollywood Westerns or the Army Corps of Engineers. With all due respect for the heroes of our youth, no other group of organisms is comparable to plants in terms of their colonizing abilities. All the more so if we include in our meaning of "pioneers" organisms capable of preparing the way for subsequent colonization by other living beings. In this sense, plants should rightfully be considered the ultimate in pioneering organisms. There is no terrestrial environment in which vegetables (understood in the largest sense of organisms capable of performing photosynthesis) are not able to take root, bringing life along with them. From the polar ice caps to the most fiery-hot deserts, from the deepest oceans to the highest mountaintops, vegetables have conquered it all, and they continue to do so every time the occasion presents itself.

I am sure that many of us have had the chance to observe—I hope with true amazement—the capacity of plants to cover in a very short time all kinds of terrain, conquering new territories or, more often, reconquering them for nature, slowly but inexorably. Years ago, not far from my laboratory at the science campus of the University of Florence, a former army depot was suddenly evacuated, as part of one of the recurring reorganizations of our armed forces, and abandoned to itself from one day to the next. The proximity of this area to my laboratory and the fact that I had studied it and lusted after it for so many years, thinking that it could become a magnificent facility for developing and trying out innovative methods of urban agriculture, allowed me to follow attentively and in detail the advance of the plants. For once, to my regret (for a long time I held on to the hope that I would actually be able to turn the area into a laboratory), I was able to see the velocity, the efficiency, and in a certain sense the strategies with which the plants reclaimed their ownership of the property. Two years after the army's evacuation, the entire perimeter wall of the barracks was covered with more than twenty different species: among them, capers (*Capparis spinose*), snapdragons (*Antirrhinum majus*), lots of spreading pellitory (*Parietaria judaica*), and several small ferns (*Asplenium ruta-muraria*).[1] All told, a little botanical garden with a vertical arrangement, and all kinds of stories to tell.

Meanwhile, at the junction between the base of the wall and the street, starting in the early months after the evacuation, a rich arboreal vegetation had begun a powerful advance. Trees of heaven (*Alianthus altissima*) and princess trees (*Pawlonia tomentosa*)—the latter certainly spawned from the seeds of a princess tree planted by me years earlier, which dominates the entire area around my laboratory and for which I have a lot of affection—sprouted up all over the place, growing in no time into powerful trees and knocking down sizable portions of the perimeter wall. A common fig (*Ficus carica*), having germinated in a crack in the asphalt of the street, is now a magnificent tree whose widespread limbs hide a sentry box carved out of the masonry wall. And then, naturally, morning glories (*Convolvulus arvenis*) arrived to cover pretty much everything, along with greater burdock (*Arctium lappa*), an inveterate hitchhiker. Today, fifteen years after the evacuation of the army depot, only a few of its structures are still able to resist the onslaught of the plants: a building made of reinforced concrete, a patio evidently capable of fighting off attacks, and, finally, an enormous metal cistern that, after having tenaciously resisted conquest for years, has recently begun to show the first signs of its upcoming surrender. In very little time, the plants have succeeded in their intent to take back an area that seemed impermeable to life. A remarkable success, but nothing compared with the legendary conquests of which plants have been the heroes.

THE PIONEERS OF SURTSEY ISLAND

In the early days of the month of November 1963, about sixty miles south of Iceland and four hundred feet below the surface of the North Atlantic, a volcanic eruption began spewing incandescent magma onto the ocean floor. At that depth, the density and pressure created by the column of water prevent volcanic emissions or any kind of explosion from being seen. As the days went by, and the accumulation of material raised the level of the seabed, the volcanic activity became more evident. From November 6 to 8 the seismic observation station at Kirkjubæjarklaustur, in Iceland (where else, with a name like that), registered a series of weak tremors, originating in an epicenter at a distance of eighty-five miles or so southwest of Reykjavik. On November 12, the inhabitants of the coastal city of Vik were disturbed throughout the day by a strong odor of hydrogen sulfide. On November 13, a fishing boat looking for herring, outfitted with excellent scientific instruments, measured the water temperature near the underwater eruption point at 4.3°F (2.4°C) higher than normal.

At 7:15 UTC on November 14, 1963, the crew of the *Ísleifur II*, cruising in those same waters, became the first eyewitnesses of the explosive eruptions.[2] They were alerted by the cook, who had seen a column of white smoke wafting up from a not precisely identified point in the middle of the sea, and changed course to provide assistance to what they believed was a ship in trouble. By 11:00 that same morning, the column of smoke and ash had risen to a height of a couple of miles and three separate volcanic vents had emerged through the surface of the water. At 3:00 in the afternoon, the three vents had fused into a single eruptive fissure. A few more days and at 63.303°N 20.605°W a new island, a little less than a mile long and 1,500 feet high, had joined the other islands in the Vestmannaeyjar archipelago.[3] The island was dubbed Surtsey, from Surtr, the fire giant in Norse mythology, who one day will return to Earth to set it afire with flames from his red-hot sword. The eruptions continued until June 5, 1967. On that date, the island reached its maximum extension, 29,000 square feet. Since then, marine erosion has constantly reduced its surface area, which in 2012 had shrunk to less than half that (14,000 square feet).

Surtsey's fate would seem to be sealed. Erosion will gradually wear it away, and, over the span of a century or so, the island will vanish from the waters where it was born. A short life, but long enough to ensure that it will remain forever in the history of science. Thanks to this rare natural laboratory, in fact, it has been possible to study, on a reduced scale and using the techniques and instruments of modern science, all the elements that contribute to the formation of a complete ecosystem from an originally sterile and inert substrate. After the lava emerged from the water and it became clear that it would not be an ephemeral phenomenon, as had happened on other occasions,[4] the scientific community outfitted itself to follow the taking root and growth of life. In 1965, when the eruptive phase was still at its height, Surtsey was declared a nature reserve and no one, with the exception of a handful of scientists, was allowed access to it. Ash, pumice, sand, and lava were waiting to be invaded by life.

It didn't take long. Plants got there straightaway, as early as the spring following the start of the eruption. In 1965, the first vascular plant, a *Cakile arctica*, or Arctic sea rocket, was growing on a sandy beach on the island. Sea rockets are surprising plants. Small, demure, not flashy, apparently devoid of interest, they are the opposite of what their aspect might make you think. Real sea wolves, rugged pioneers, present at all latitudes, these plants that live along the coasts are capable of undertaking long sea voyages and surviving without any source of fresh water. All species belonging to the *Cakile* genus, in fact, are halophytes (from the Greek *halas*, "salt," and *phyton*, "plant"), that is, endowed with particular modifications, both anatomical and physiological, that make them able to grow using seawater, in conditions impossible for the survival of other species.[5]

But that's not all. Evolution has been lavish with sea rockets, providing them with a survival kit they can use whenever the need arises. A little like the souped-up Aston Martins in James Bond movies, sea rockets can put into play an arsenal of tricks designed to ensure their maximum performance on all occasions. One of my favorites is the extraordinary way these plants have of dispersing their seeds. When the seeds are ripe, the pod that contains them breaks in two. Half falls to the ground close to the mother plant, burying itself in the sand and ensuring that in any case some seeds will have a good chance to germinate.[6] The other half is swept out to sea. Here, the seeds, endowed with an excellent capacity to float, can remain vital for years, until the marine currents deposit them on some distant beach for them to spread themselves on. That's how, in the race to see who would be the first to get to Surtsey Island, the sea rocket managed to outclass all the other contestants.[7]

The work of census-taking that followed the colonization of Surtsey produced some surprising results right from the start. For example, no one expected that one of the carriers by which seeds got to the island might be fish eggs. To be more precise, the typical capsules that contain skate (*Raja batis*) eggs, transported, as unexpected guests, seeds of various species of herbaceous plants. Apart from this original means of transportation, most of the seeds reached the island transported by wind, water, or birds. Snow buntings (*Plectrophenax nivalis*), for example, cute little sparrows that love rather bracing climates, contributed actively to the spread of seeds to the island as they migrated to Iceland from Scotland, transporting in their own little gizzards (the food-grinding stomachs of birds) seeds that passed unharmed through their digestive apparatus and happily managed to germinate. This was the route by which plants such as *Polygonum maculosa* (a lovely cosmopolitan shrub) and *Carex nigra*[8] (a wetlands grass also known as smooth black sedge) came to the island as early as 1967. Even marine birds, such as seagulls, which do not commonly feed on vegetable matter, actively contributed to the arrival of new plant species by sometimes taking nourishment from plants in distant arid zones and transporting the seeds to the island.

Finally, geese, by dropping their excrement from on high as they passed over Surtsey, proved to be exceptional carriers, capable of depositing on the island a vast variety of seeds coated with natural fertilizer and, therefore, in the best possible condition to germinate.

Of all the vascular plant species recorded on the island, 9 percent were transported through the air by wind, 27 percent by sea, and the remaining 64 percent by birds.[9] At the end of 1998, the first exemplar of an arboreal species, a *Salix phylicifolia* (tea-leaved willow), finally took root on the island. In 2008, forty-five years after its birth, sixty-nine species of plants had been recorded on Surtsey, of which thirty were by now considered settled. Still today, other species continue to arrive, at the rate of two to five per year.

THE COMBATANTS OF CHERNOBYL

The Chernobyl disaster is one of those catastrophes that have a permanent place in human memory. I imagine that, even among my youngest readers, there are not many who are unfamiliar with what happened there. Nevertheless, just for the sake of leaving no one in doubt and to refresh the memory of all the others, here is a quick summary of the known facts.

At 1:23 a.m. (local time) on April 26, 1986, nuclear reactor number four of the nuclear power plant Vladimir Ilyich Lenin, located just over eleven miles from the city of Chernobyl in Ukraine (at the time still part of the Soviet Union), exploded. The explosion was caused by a series of contributing factors, attributable to important construction defects and unprecedented negligence on the part of technical personnel, who violated numerous safety protocols. Following an error during some tests, the sudden rise in temperature inside the reactor core brought about the splitting of water molecules into hydrogen and oxygen. The inevitable contact between the hydrogen and the incandescent graphite of the control rods (which serve to regulate the nuclear fission reaction) provoked an enormous explosion. An explosion so strong as to destroy the reactor casing and blow its hermetically sealed covering plate, weighing more than a thousand tons, through the roof of the building. The fire that followed dispersed into the atmosphere an incredible quantity of radioactive isotopes, which were deposited in the area immediately surrounding the power plant and, in part, riding on the high-altitude wind currents, made it to western Europe (except for Spain and Portugal) and North America.

It was the first level seven—the most dangerous—nuclear accident in history. The second would be the accident at the nuclear power plant in Fukushima, Japan, on March 11, 2011. There were fifty-seven immediate victims of the Chernobyl catastrophe, but the number of people exposed to the radioactive isotopes who developed a fatal pathology in the ensuing years would reach tens of thousands. Arriving at an exact number is difficult; estimates range from the thirty thousand to sixty thousand of the United Nations official report to the more than six million of Greenpeace. Because of the accident, the entire city of Chernobyl and a vast area surrounding the power plant were completely evacuated, and more than 350,000 people had to be "resettled" in other regions of the Soviet Union. The evacuated area, the so-called Zone of Alienation, covering a radius of eighteen miles around the plant, was completely sealed off and access was denied to everyone for decades.

The effects of the Chernobyl disaster were so devastating that still today, more than thirty years after the accident, we have only a vague idea of the consequences of what happened and of how long we will have to wait until things return to normal.

Plants, too, obviously, were exposed to radioactive fallout during the days following the explosion, and for them, too, the consequences were catastrophic. It has been calculated that in the first weeks after the accident, 60 to 70 percent of the radioactive isotopes released into the environment were deposited on the plants and trees in the surrounding forests. A considerable part of those Scots pine forests, including those in the Zone of Alienation, died immediately, changing color from green to red and giving rise to the phenomenon known from then on as "red forest." In 2011, the radioactive fallout after the accident in Fukushima provoked the same phenomenon.

After the dramatic effect of their initial exposure to the high doses of radioactivity had come to an end, the plants found a way to survive and adapt even to these conditions, seemingly incompatible with life.

What happened inside the Zone of Alienation is unbelievable. This still-inaccessible space for people has become one of the most biologically diverse territories in the former Soviet Union. It seems that humans are much more harmful than radiation. The exclusion of human activity from these areas has in fact created an enormous involuntary nature reserve. Notwithstanding the radiation, plants and animals have returned in superior numbers and varieties than in the past. Today, in the Zone of Alienation, there are lynx, raccoons, roe deer, wolves, Przewalski's horses, various species of birds, elk, red foxes, badgers, weasels, hare, squirrels, and even brown bear, which had disappeared from the area more than a century ago.

And what about the plants? Obviously, they did much more and much better than the animals. The city of Pripyat, situated within the Zone of Alienation, was just under two miles away from the exploded reactor. It was a city of around fifty thousand inhabitants, where most of the workers at the power plant lived. After the accident, it was completely evacuated.[10] I have recently had the opportunity to see a highly detailed photographic report on the state of the city today. The images are stunning. Thirty years after the disaster, Pripyat is covered with plants—a veritable Ukrainian Angkor Wat. Poplars on rooftops, birch trees on terraces and balconies, asphalt split open by shrubs, enormous six-lane highways transformed into rivers of green.

The response of plants to the Chernobyl disaster was so unexpected that even experts in the field were astounded. Unfortunately, despite a lot of interest in the phenomenon, serious scientific studies are almost nonexistent. In 2009, a team from the Slovak Academy of Sciences, led by Professor Martin Hajduch, went all the way inside the city of Pripyat for an experiment whose results produced a heated discussion. The team disseminated throughout the city a certain quantity of flax and compared its growth and productive performance with those of an equivalent group of plants cultivated more than sixty miles from the contaminated area. The result was that the plants inside the city of Pripyat grew much larger than the others while

Crithmum
Harbour

Hogetory Sea

Calendula Island

Honaalaa City

Philodendrum City

Schinus

Capobraty Beach Cap Tagmoinoides

Last

consuming proportionately less water. The article reporting on the experiment attributed this result to a series of proteins, which, present in greater quantities in the plants grown in the contaminated area, may have protected them from the damaging effects of the radiation.[11]

Although these results are subject to criticism, related in part to the difficulty of comparing growth in such different places (quite apart from the radiation), it is sure that, over the course of their history, plants have developed an extraordinary capacity to stand up to adversity.

It is well known that one of the most astounding capabilities of plants is their capacity to absorb radionuclides, subtracting them from the environment. Many plants succeed in this apparently impossible enterprise, and their use in cleansing the environment of these contaminants through a technique called phytoremediation has often been proposed.[12] Although it is not extremely fast, this technique offers the only real possibility of decontaminating terrain polluted by radionuclides. Every other method requires the movement of earth with the consequent production of dust and concomitant risks that strongly advise against their use. The amount of radioactive material absorbed can vary considerably, depending on climate, terrain, soil composition, and so on.

The fact remains that over time, plants, by absorbing radioactive material, remove it from the environment, concentrating it in themselves. That is what is happening in the Exclusion Zone at Chernobyl. Obviously, this poses some very serious problems. What would happen, in fact, if a fire swept through those forests? The radioactive material accumulated in the trees and plants during the last thirty years would release immediately into the atmosphere with grave consequences. That is why fire prevention in the evacuated zone is one of the priorities of the Ukrainian government.

THE *HIBAKUJUMOKU*, OR THE VETERANS OF THE ATOMIC BOMB

I didn't know about the existence of the *Hibakujumoku*; I came to find out about them by pure coincidence a few years ago, during one of my periodic visits to Kitakyushu, in Japan. This city, which hosts an office of the LINV (International Laboratory of Plant Neurobiology)[13] directed by my friend Professor Tomonori Kawano, has been for years my personal port of entry to Japan and its culture. Every time I go there, I try to carve myself out a little free time to get to know something new about this distant empire. One of the activities I enjoy most is having lunch or dinner on my own in some typical local eatery, though I know almost no Japanese, except for some very simple polite phrases . . . and the numbers, written and spoken.

Japanese cuisine is so varied and refined that it's hard to happen upon something unpleasant to the palate. My personal procedure, therefore, is to take a seat at the counter and start pointing, completely at random, to a series of dishes chosen on the basis of how much I like the characters with which they are represented. Normally, they turn out to be small dishes, with small servings, that in no time start crowding the counter space allotted to me, turning it into a little work of art. That's the moment I like best: you feel the rush that comes with gambling, but without the risk, except the negligible one of a truly unsavory dish. Then comes the pleasure of discovery: What am I eating? What are the ingredients? How was it prepared?

During one of these blind dinners, I happened onto a mysterious dish that defied all my efforts to figure it out. It was a sort of whitish little pouch about the size of a ravioli, lightly fried and filled with a creamy substance tasting of fish. The flavor was delicious and so, having finished the first serving, I promptly ordered a second, so as to study it in depth. It reminded me of something from Italian cooking but I couldn't put my finger on it. I puzzled over it for a while but nothing came of it. I even tried to talk to the waiter about it, but in Japan hardly anyone speaks anything but Japanese. Disconsolate, I was ready to ingest the second serving, leaving my queries unanswered, when something totally unexpected happened—one of those things that make me love going out to eat by myself in Japan. An elderly man, sitting next to me at the counter, addressed me. That in itself is crazy. Never in all my years of visiting the Land of the Rising Sun had anyone spoken to me without being spoken to first. I had always been the one to start a dialogue. But that's not all: he spoke to me in perfect, elegant Italian that hit a bump just for a second, right at the start of our conversation, when, embarrassed, he couldn't find the right words to tell me, without scaring me, what I was eating.

"You see, dear sir, we give great importance to the reproduction of life," he began, leaving me a bit disoriented. "Even though our culture is often noted in the West for its aggressive connotations, in reality there is a strong component of panpsychism (that's the word he used) in our civilization." I reassured him, for reasons of courtesy, that this aggressive connotation of Japan belonged more to the past than to the present. The look on my face, however, must have remained puzzled. What did panpsychism have to do with what I was eating? He tried again.

"As a consequence of the presence of the divinity in everything, we tend by tradition to consume every single part of animals." Now we seemed to be getting closer. And so?

"So, that dish that you are consuming, which by the way is called *shirako* and is without a doubt among my favorites, too, is produced from the male germinal line of various marine species."

"The male germinal line? You mean . . ."

"Yes, what do you call it in Italian?"

"Sperm."

"Exactly."

So that's what it reminded me of: milt, an exquisite Sicilian specialty with the sperm sacs of tuna or *ricciola* (greater amberjack), the male equivalent of bottarga or roe. The fact that in Italy people eat the same not-so-noble parts of fish (I believe for reasons much more material than those associated with panpsychism) reassured my new dinner companion. We introduced ourselves; he was a retired diplomat. During his career, he had served his country as consul in Italy for many years, and he had learned the language. We went on to talk for a long time and with great pleasure until, just before saying good-bye, he asked me: "I imagine, Professor, that you have already had occasion to visit our *Hibakujumoku*," and left the question floating in the air between us for a few seconds. I answered that I had never heard of them and that I was sorry about that. Whatever the *Hibakujumoku* might have been, it is not polite in Japan to say you don't know about something without excusing yourself. The consul was very struck by this knowledge gap of mine.

"But you are concerned with plants! You simply must meet them." He said exactly that, "meet them," so I thought he was referring to a group of people who in some way were concerned with plants. But his next words shattered my supposition. "The *Hibakujumoku* are our escapees from the atomic bomb. A living hymn to the force of life." I knew that in Japan survivors of the bomb attacks in Hiroshima and Nagasaki held a fundamental position as witnesses to those atrocities. But I couldn't understand the reason he was so insistent that I should meet them. The mystery did not last long. "They are not people, but trees exposed to the atomic bomb.[14] In Japan, everyone knows them and respects them. Personally, I love them. You should know them, too. I am going to allow myself to make a proposal. Hiroshima is not more than two hours from here by train. If you like, I can accompany you there in the next few days to meet them. Please tell me if you would like to do that. Since my wife died, my days are mostly free of commitments." I thanked him intensely and accepted his offer with pleasure.

Two days later, each of us armed with his own *bento*,[15] as befits two friends on an outing, we met early in the morning in front of the train station in Kokura, ready for our trip to Hiroshima. In a little over an hour and a half, we arrived at the station in Hiroshima, and ten minutes later we were standing before my first *Hibakujumoku*. The consul had led me through a magnificent garden—whose name I unfortunately don't remember—to "meet" three trees that had survived the bomb. I remember them very well: a ginkgo (*Ginkgo biloba*), a Japanese black pine (*Pinus thunbergii*), and a *muku* (*Aphananthe aspera*), three very common trees in any classical Japanese garden. The ginkgo was conspicuously bent in the direction of the city center, the black pine had a considerable scar on its trunk, but all in all, they were in good shape. Normal trees by all appearances, if it hadn't been for the evident feeling of respect and, I would say, affection that they inspired in the people who were there to "meet them." A genteel elderly couple (probably husband and wife) had taken a seat on two portable chairs in front of the ginkgo and were engaged in a long conversation with the tree. A young boy gave it a quick hug before continuing on his walk. Everyone who passed by the trees seemed to know them well, and many people, from kids to old folk, bowed deeply before them. On each *Hibakujumoku* hung a yellow sign, the only characteristic that distinguished them from the other trees. I asked the consul what it said.

"I'll try to translate it for you. It says more or less that we are standing before a tree that suffered an atomic bomb attack. Then it gives the vegetable species and finally the distance from ground zero," he explained, pointing toward the river. "The explosion happened down there, where the river forks, exactly 4,494 feet from here."

That day I visited a lot of *Hibakujumoku*, making my way closer and closer to the place where, for the first time, an atomic ordnance was used against a defenseless population. I remember another magnificent ginkgo inside the enclosure of the Hosenbo

Temple at 3,700 feet. A camphor tree (*Cinnamomum camphora*) inside the quadrilateral of the Hiroshima Castle, at 3,674 feet. A Kurogane holly tree (*Ilex rotunda*) also inside the castle grounds at 2,985 feet. A marvelous peony (*Paeonia suffruticosa*) in the temple of Honkyoji at 2,920 feet.

As we moved in closer to the center of the disaster, the *Hibakujumoku* began to thin out. At 8:15 in the morning on August 6, 1945, the ground temperature in the place where we now were had risen above 7,200°F (4,000°C); very probably it went as high as 10,800°F (6,000°C). The consul had just taken me to see the shadow (literally) impressed on the stairway of the Sumitomo Banking Corporation, left by the vaporization of Mrs. Mitsuno Ochi, age forty-two at the time, caught unawares by the explosion as she was waiting for the bank to open. No hope that anything could have survived all that destruction. I voiced this thought to the consul, who responded, smiling, "Man of little faith. Life always wins! Follow me." We went around the corner and found ourselves once again along the Honkawa River. The "atomic bomb dome," the only building left standing and preserved as a peace museum, which by convention marks ground zero, was there in front of me, at less than 1,300 feet from where we were, and there, too, right in front of us on the riverbank stood the champion of the *Hibakujumoku*, a weeping willow (*Salix babilonica*) reborn from its roots left alive underground. Its yellow sign indicated a distance of 1,214 feet from ground zero.

On the way back home to Kitakyushu that evening, the consul invited me to dinner at an inn he knew. I gladly accepted. It was a very pleasant evening and we drank a lot, as often happens among friends in Japan. As we spoke about our "encounters"

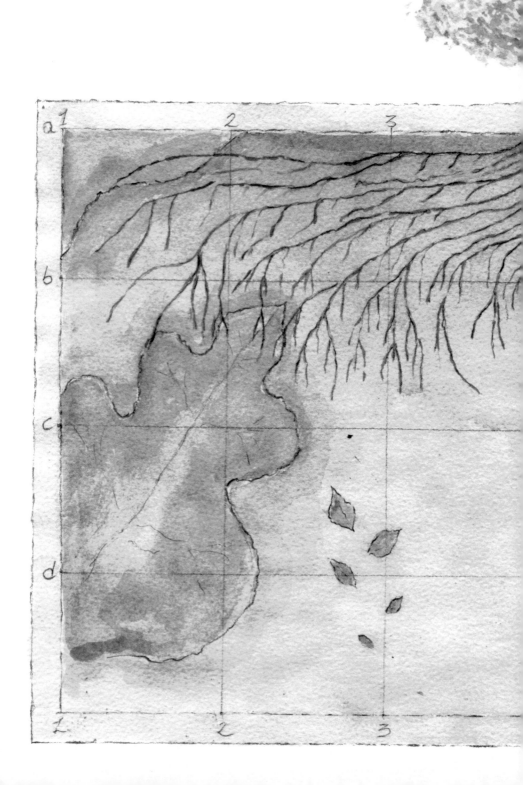

in Hiroshima, something left me puzzled. Each time the consul spoke of the *Hibakujumoku*, he defined them as "trees that had suffered an atomic explosion," and this long circumlocution sounded funny and strangely discordant with his otherwise perfect proficiency in Italian. So I dared to venture, "Excuse me, consul, why do you keep saying that the *Hibakujumoku* are "trees that have suffered an atomic explosion"? Wouldn't it be simpler to use a word like "survivors"?

Here is his explanation: "The question is more complicated than it seems, dear Professor. It all starts with the name given to the survivors, as you say, of the bomb. Their name in Japanese is *hibakusha*, literally 'person exposed to the bomb.' There is a reason for this choice that you can grasp. This term was chosen rather than 'survivors' because that word, by exalting those who had remained alive, would have inevitably offended the many who died in the tragedy. As a consequence, the *Hibakujumoku* are referred to in the same way. I imagine that seems strange to you, but I assure you that all *hibakusha* are content about this and could not have abided being called 'survivors.'" I suggested the Italian word *reduci*, or "veterans." He didn't know it and he liked it very much. "Thank you so much for teaching me that word. It sounds very good. Let's toast to our veteran friends."

After leaving the restaurant, I insisted on accompanying him home. No one would have guessed it from looking at him, but the consul was well over eighty years old, and he had had a lot to drink. In any event, he agreed, and I accompanied him on the short walk to his home. We said good-bye. Breaking every Japanese rule, by virtue of his many years spent in Italy, the consul hugged me. He looked me straight in the eye and said,

"Talk about the *Hibakujumoku*, make them known. And come back to visit them again." Then he paused, trying to decide. "I have to tell you. I, too, am a *hibakusha*. I was seven years old when the bomb did away with my whole family and everyone else in the world I knew. I was saved because the classroom in the elementary school where I was studying was protected by a curtain of trees. I and four of my classmates are the only veterans from that school. We were one hundred twenty children."

He thought about that for a second, smiled at me one last time, and, turning to pass through the door to his home, thanked me again for my company.

Altea rosita   Ambienti   Asfodelo

Abelia

Echeveria
Crassula

Arbutus Tosun

Channel

Lagerstromia

Angelops

Aucuba

Chamaerops

Alium

Cupressus
Harbour

Lotus

Gaussia
City

Spinosa

Bazzet

Lepidio

Port Ferula

Reserve

Tempest Island

Anacamplis

Aloe
Bay

Asterisco

Judica

Ayezato

Ficus Beach

Calendola
Cap

Agazzina

Agerato   Acacia Saligna   Adonide annua

. 2 .

FUGITIVES AND CONQUERORS

type species Pennisetum – domain Eukaryota – kingdom Plantae

division Magnoliophyta – class Liliopsida – order Poales

family Poaceae – subfamily Panicoideae – tribe Paniceae

genus *Pennisetum* – species *Pennisetum setaceum*

origin North Africa – diffusion Worldwide

first appearance in Europe Twentieth century

 The expansive thrust of life cannot be contained. This makes it impossible to keep a vegetable species enclosed inside fenced-off areas, such as botanical gardens or nurseries. Despite our constantly trying to do just that, sooner or later, the plants manage to escape, winning back the chance to continue their expansion.

Most species, whether animal or vegetable, that we consider today to be invasive came to be so in this way: by escaping from places where people believed it was possible to keep them confined. To be more precise, not only species we consider invasive but most of the plants we think of as having always been a part of our environment are actually migrants from a more or less distant past. Plants that are now perceived as part of our cultural heritage are merely well-assimilated foreigners.

Take corn. This foreigner from Mexico[16] has fed the population of the Po Valley for generations. Or tomatoes and basil, distinctive plants of Italian food culture—after all, isn't pasta with tomato and a sprig of basil Italy's national dish? Well, the tomato (*Solanum lycopersicum*) is an original species of an area that extends from Mexico to Peru, brought to Europe for the first time by Hernán Cortés in 1540. And it was still nothing like the tomato we know today. Indeed, when it got to Italy in 1544, its fruit was yellow and it was described by Andrea Mattioli in his *Medici Senensis Commentarii* as *"mala aurea,"* later translated more literally as *"pomo d'oro"* (golden apple). In order to gain acceptance, the poor tomato, as happens with many other migrants, had to go through what might be called a "multicolored experience." And in the case of the tomato, the phrase must be taken literally. Indeed, until it turned red it was looked upon with suspicion. First, it was considered toxic, then of only ornamental use, and then curative. Not until 1572 do we find a reference to a "lustily red" tomato. From then on, everything got easier; once it veered to red, the tomato was almost home. It began to be used for nutritional purposes. But slowly. So slowly that the first recipe for our national dish, pasta with tomato, would not appear until the first half of the nineteenth century.

A long journey, but all things considered a simple one compared with the experience of basil, that other bulwark of Italian cuisine. Basil (*Ocimum basilicum*) is a foreigner, too. It comes from the interior of India and was brought to Europe by Alexander the Great. By comparison, the tomato was welcomed with open arms. Before taking its place on our dinner tables, basil had to wait from 350 BCE until the eighteenth century. For more than two thousand years, our aromatic stranger enjoyed a very bad reputation; from Pliny the Elder, whose *Natural History* held that basil was responsible for states of torpor and madness, all the way to Nicholas Culpeper, the British physician and botanist from the first half of the seventeenth century, who deemed it neither more nor less than a poison.[17]

But now, let's leave aside food plants and all other plants that were introduced in order to be put to use in some way. For them, economic or utilitarian analyses have always prevailed over any kind of naturalistic consideration. What is interesting to highlight is that, apart from cultivated species, many plants that we look on today as part of our native flora are not native at all, having originated in places that are often very far away indeed.

Why, then, do we insist on labeling as "invasive" all those plants that, with great success, have managed to occupy new territories? On a closer look, the invasive plants of today are the native flora of the future, just as the invasive species of the past are a fundamental part of our ecosystem today. I would like this concept to be perfectly clear: the species that we consider invasive today are the natives of tomorrow. Keeping this rule firmly in mind would prevent many of our ridiculous attempts to limit their expansion.

The qualities that make plants invasive are numerous.[18] Let's recall some of them: great ability to spread their seeds;

rapid growth; the capacity to alter their form in response to environmental conditions;[19] tolerance of multiple kinds of stress; capability of associating with humans. All in all, these are the characteristics that make a species efficient, flexible, and resistant, capable of resolving all the problems that each new environmental situation might pose. In short, they are the qualities that describe intelligence. I have no doubt about that. That's why species that have the capacity to adapt to new environments are the species I love best. They are the most interesting and the ones that have tricks that are worth knowing. In the pages that follow, we will be talking about three irresistible fugitives.

FROM ISLAND TO ISLAND

Senecio squalidus. Although its name doesn't seem so attractive, it is an elegant and gracious plant, belonging to the family of *Compositae* or *Asteraceae*, the family of flowering plants that boasts the highest number of species: 32,913, divided into 1,911 genera.[20] The genus *Senecio* alone, to which the protagonist of our story belongs, groups together more than a thousand different vegetable species. The genus name *Senecio*, from the Latin *senex*, old or aged, refers to the characteristic pappus,[21] or tuft of thin, spindly whitish hairs, all the same length. The species name *squalidus*, on the other hand, we owe to Linneaus, who so named it in his *Species plantarum* of 1753.

This species came into the world on the slopes of Mount Etna. A hybrid, born probably from the union of *Senecio aethnensis* and

Senecio chrysanthemifolius,[22] it has its own undeniable grace. From twelve to twenty inches tall, with lanceolate leaves and pretty yellow flowers clustered in corymbs, this plant has managed— starting out from the slopes of Etna—to conquer all of Great Britain. Understanding how it did so is of the utmost importance. Indeed, the study of its uncontainable expansive thrust allows us to discover the important characteristics that determined its expansion. We can measure, for example, the pace of its diffusion, whether there was—as is quite common—an initial phase of latency, its effects on local flora and fauna, and so on, but, above all, whether an evolutionary change in the populations took place and, if so, what the traits were that favored its diffusion. It is not surprising, therefore, that our little plant, catapulted into Great Britain from Sicily, and for which there exist detailed records, has become a model for the study of these phenomena.

The first to note and describe it was a Sicilian botanist from the second half of the 1600s, Francesco Cupani, a Franciscan friar trained in the school of another Sicilian friar, the Cistercian monk Paolo Boccone, from Palermo. A professor of botany in Padua and then botanist at the court of Grand Duke Ferdinand II of Tuscany, Boccone was one of the great advocates of the need to modernize botanical taxonomy. With such a teacher, it was inevitable that Francesco Cupani should also cultivate, for the greater glory of God, an unconditional love for botany. A love so strong, in fact, that it inspired him to take up the gigantic and complex enterprise of cataloging and describing the entire flora of Sicily, one of the richest in Europe. In 1692, in order to bring this task to completion and have a place where he could conserve living exemplars of the species he classified, Cupani,

Liliaceae beach

Juncus

Juncandaceae

Cap
Gasteria

Abutilon
cap

N
aziis

NO

NE

O

E

SO

S

SE

Datuzu Sea

Guzde

Bruguiania

Dieffenbachia

Dasyliriou
city

Dracena
Terminalij

Ginkgo
Peninsula

Dahlia
Gulf

1,35"   4,27'   6,22'   8,17'   H.

h,i"

9,i"

y°x'

15°1"   11°1"   12.6"   17°0°

12°,7'
14°,2'
16°,-7'
y°x'
Zizyphus Gulf
Albus
Hyoscyamus
Isatis
Patagonium
l°,m'
Z°,9'
Grevillea Bay
Gladiolus promontory
Little Seeds
Archipelago
Agave
Islands
Guaderella
Cap
quayabo
X°,1'
Dafne Land
Grevillea

with economic support from the Duke of Misilmeri, Giuseppe del Bosco Sandoval, founded a botanical garden in Misilmeri, not far from Palermo.[23] In Cupani's all-too-brief lifetime, his botanical garden became famous throughout Europe.

In his endless search for all Sicilian species, it was inevitable that Francesco Cupani would finally also happen upon our *Senecio squalidus*. As was the case for all the species he collected, this one was also transferred and propagated in the botanical garden in Misilmeri. There, Cupani planted Sicilian species next to other species from all over the world and classified them according to a binomial nomenclature, a forerunner of the system that would be more generally adopted only many years later, thanks to the work of Linnaeus. Like all botanical gardens worthy of the name, the Misilmeri garden, in the attempt to enrich its collection as much as possible, undertook an energetic campaign of relationships and exchanges with other botanical gardens of the time. It is reasonable to suppose—but this is only a hypothesis of mine unsupported by documentary evidence—that in the course of pursuing these relations with foreign gardens Francesco Cupani came to know William Sherard, an illustrious English botanist, making him a present of some *Senecio squalidus* seeds for his English collections. However things actually went, in the year 1700, during the same period in which Sherard was serving as tutor for the Duchess of Beaufort, *Senecio squalidus* was happily installed in the family's ducal gardens at Badminton. A few years, or maybe only a few months, later, our Sicilian hybrid was introduced by the *horti praefectus* Jacob Bobart the Younger into the botanical garden at Oxford, the launching pad for its invasion of Great Britain.

Senecio squalidus, you will recall, originated in the volcanic ash and lava on the slopes of Mount Etna. It is, therefore, a very rugged plant and adept at living on scarce resources. Its favorite urban habitats are city walls, ruins, and courtyards. In other words, all those places that other vegetable species are not fond of in the least. In just a short time, our Sicilian lady was a well-known resident of Oxford. In 1794 not a single college wall was without its exemplars of *Senecio squalidus*. Its little yellow flowers sprang up even on the walls of the Bodleian Library, the very symbol of Oxford. Finally, the plant was officially adopted by Oxford, which gave it the name by which it is commonly known in English: "Oxford ragwort." The first exemplars began to spread in the vicinity of the city, taking possession of derelict farms and the walls of abandoned buildings, the first step in its gradual expansion toward the rest of England. The conquest proceeded slowly until the arrival of the railroad brought a decisive change in the pace of the invasion.

On June 12, 1844, the Great Western Railway inaugurated its station in Oxford, connecting it to London. In the following years, other rail lines brought Oxford into communication with the rest of Great Britain. Our *Senecio* adapted marvelously to the magnificent progress of the Industrial Revolution and became one of the first and most enthusiastic riders of the rails. The gravel beds between and along the rails, whose purpose was to keep plants from growing there, turned out to be irresistible. They were an intense reminder of the lava, ash, and sand in which she sprouted in her by-now-distant native land. Our little lady suddenly found herself in a perfect situation with no obstacles to rapid propagation. The excellent substrate in which to grow was joined by the helping hand for diffusion offered by the frequently passing trains. All *Senecio*, in fact, thanks to the white filaments that gave rise to their name (remember the pappus?), make use of air currents to spread their seeds. Over the course of a year, the plant can produce huge numbers of fruit, whose seeds, transported by the air set in motion by each passing train, are ready to propagate. Yard by yard, following the railroad tracks, our Sicilian lady set off to conquer the north of Great Britain. Toward the end of the nineteenth century, it had reached numerous locations in northern England; in the 1950s it arrived in central Scotland, and then went on to spread to the north of Scotland and across the North Channel into Northern Ireland, benefiting in recent years from the traffic on the motorways.

Making use of human transport (trains and automobiles) as a propellant for the flight of its seeds gave *Senecio* a further advantage. Imagine for the moment that a seed, carried for some reason to some distant region, manages to make a home for itself far away from its own population of origin. Its chances of consolidating its presence in this new territory will be extremely scarce. At low population densities, in fact, it is difficult for plants to reproduce and thus to spread in a new place. Normally, it is necessary for a species to make repeated forays into the same place before it manages to establish itself there. The recurring sequential movements provoked by the continuous passage of cars and trains over the same roads and rail lines offered the *Senecio* exactly the kind of repeated opportunity it needed in order to settle in.

As unstoppable as the Golden Horde—documents in the Biological Records Centre provide a graphic demonstration of the conquest—*Senecio* occupied all of Great Britain. Nevertheless, its advance presented a mysterious and inexplicable trait. Notwithstanding its ability and capacity to turn circumstances to its own advantage, how is it that a species originating in Sicily managed to withstand the climate and environment of Scotland and Ireland? It didn't take long for the mystery to be unveiled. As it journeyed north, the plant learned to hybridize with local species. As long as the hybrids are not sterile, this strategy is brilliant. By crossing itself with local populations, *Senecio* gave rise to a series of hybrids that natural selection could go to work on. In this way, it quickly acquired all the genetic elements it needed to adapt to its new environmental conditions. From that moment on, *Senecio* was no longer Sicilian but Anglo-Sicilian. Following the example of other conquering dynasties, she became a naturalized Briton, an integral part of her new environment. The invasive species of yesterday became the native species of today. As was to be demonstrated.

BEAUTIFUL ABYSSINIAN

Another fugitive with a fascinating history is a little Abyssinian migrant known as *Pennisetum setaceum*. This time the Latin moniker refers to a perennial grass just over three feet tall that produces a lovely, soft and feathery spikelet (hence the name *setaceum*, or "bristle"). Initially dark pink, during ripening the spikelet gradually fades through the various shades of pink, transforming the plant into a delicate symphony of colors. While the *Senecio squalidus* might not seem as gracious to everyone as it does to me, the beauty of the elongated crimson fountain grass (common name) is plain to see and universally recognized,[24] to the point that the species is cultivated as an ornamental plant all over the world. It could even be said that her alluring aspect has been the Trojan horse by way of which she has managed to spread to any place where the climatic conditions are in harmony with her status as a daughter of sub-Saharan Africa.

In some respects, the crimson fountain's story of flight and conquest resembles that of *Senecio squalidus*, starting with the theater of operations: Sicily. However, while for *Senecio squalidus* the island is the region of origin, for the crimson fountain it is the region of conquest.

Pennisetum setaceum arrived in Sicily in 1938,[25] thanks to the interest of Professor Bruno, dean of the School of Agriculture at the University of Palermo, who, having acquired a sample of seeds, sowed them in the colonial garden attached to the botanical garden in Palermo. He started to study the plant's characteristics of growth and production, with an eye toward its potential use as forage for farm animals. Environmental conditions in Abyssinia, then an Italian colony, are not unlike those in Sicily, Professor Bruno reasoned. If this little plant showed itself capable of adapting to the island environment, it could become an excellent forage plant suited to hot, dry climates.

Unfortunately, despite adapting magnificently to its new environment, the plant had a very low nutritional value, and, what's more, animals didn't seem to like it. His hopes of turning it into a forage plant having been dashed, and there being no other practical interest in keeping it, Bruno decided to eliminate it from the colonial garden to make room for new experiments. This is where the plant's alluring aspect came to the rescue. Having noted the beauty of its flowering, the technicians who worked in the botanical garden decided to keep it in cultivation and exploit its potential as an ornamental plant.

This narrow escape was an alarm bell. There was no more time for indecision—the crimson fountain had to speed up its preparations for breaking out of the garden where it had been confined. Indeed, although Sicily did not seem to be enchanted with the *Pennisetum*, the obverse was not true at all; the plant liked the island a lot. It offered an environment similar to its native land but without all of its natural enemies and rivals in the daily struggle for survival. She decided to accelerate the pace of her diffusion.

Beauty certainly helps in the work of territorial expansion, but by itself it was not enough to satisfy the ambitions of our little friend. Very wisely, she made use of other weapons in order to extend her foothold in this new environment. Let's take a look, then, at some of the weapons that make crimson fountain grass one of the fastest-spreading species we know. First of all, she adapts to very diverse climates. As long as it rains less than five inches per year and the temperature doesn't drop below 32°F, everything is fine. She reaches sexual maturity in the second year of life, and, from then on, her floral production in the Sicilian climate is practically nonstop, from March through September. Furthermore, she stands up extremely well to drought and high temperatures and is perfectly adapted to the passage of ground fire. Thanks to this last ability, the species reestablishes itself on scorched terrain better than its direct Sicilian competitors.

Her seed, moreover, is a marvel. It knows no dormancy;
in optimal conditions it germinates immediately. If, on the
other hand, conditions are unfavorable, it is able to maintain its
vitality in the soil for up to six years. Seed production is high,
and dispersal can be accomplished by all kinds of carriers: wind,
water, animals, people, and, above all, vehicles. Thanks to roads
and the vehicles that use them, in fact, the crimson fountain
would go on to conquer Sicily.

But let's not get ahead of ourselves. One thing at a time.
The first step is to get away from the cramped flower beds
where the *Pennisetum* is confined in the colonial garden.
The escape is child's play. A windy day, of which there
are plenty in Palermo, is the ideal situation for thousands
and thousands of feathery seeds, designed to propagate by
air, to take flight. Having sailed over the garden walls and
landed happily on the abandoned area of the flower beds
immediately outside the botanical garden, the conquest can
begin. Once she's on the outside it's only a question of time.
Using the uncultivated areas on roadsides and exploiting a
technique similar to that of the *Senecio squalidus* in Great
Britain, the *Pennisetum* begins to spread, following the main
roads heading out of Palermo. There are maps showing its
itinerary in Sicily in recent decades, and they are uncanny;
the plant's advance throughout the island is an exact tracing
of the road network. Every year the plant propagates itself
by conquering dozens of miles, and today practically all of
Sicily is home to crimson fountain grass. A Sicilian species
conquers Great Britain and an Eritrean species conquers
Sicily. True globalization. It has existed forever in nature. For

plants, fortunately, tariffs, borders, travel bans, and barriers are meaningless concepts.

HIPPOPOTAMI IN LOUISIANA

One plant that truly has a terrible reputation in many parts of the world, and with all of the national and international agencies involved in some way with invasive plants, is without a doubt the *Eichhornia crassipes*, or water hyacinth. Its rapid diffusion and its sovereign contempt for the vast majority of means with which humanity tries to fight it have combined to make it commonly considered the worst aquatic invasive species known to humanity. Furthermore, it has the dubious privilege of membership in the elite club of the "100 worst invasive species" established by the Invasive Species Study Group (ISSG).[26] In short, deemed the vegetable personification of evil, it is hated by everyone. Without reservation. As you might imagine, it is exactly the kind of flora non grata that I find irresistible.

First of all, I wish you could see it. You could never imagine that a monster of this sort could hide behind such a delicate and lovely appearance. The water hyacinth is an aquatic plant with origins in the Amazon, capable of floating thanks to its bulbous, spongy stems, which retain large quantities of air. It has large, shiny, thick leaves, which can form a stratum of vegetable matter on the surface of the water going down as far as three feet deep. Its beautiful and numerous flowers range in color from lavender to pink. They are the bait that has made this plant irresistible. Up until the end of the eighteenth century, in fact, the species was appreciated for its decorative qualities, for which it was imported to Europe. Today, it is present in more than fifty countries on five continents.

The success of this plant, as we were saying, was initially tied to its beauty. First identified and described in 1823—the genus *Eichhornia* is named after the Prussian prime minister Johann Albert Friedrich Eichhorn (1779–1856)—the species spread quickly throughout the world by using botanists and their botanical gardens as its passkey to the most distant corners of the planet. In repeated campaigns of conquest, the water hyacinth reached every tropical area in the world, spreading outward from Europe, where it arrived in strength, inhabiting public and private gardens throughout half of the continent, in the second half of the nineteenth century. From here, thanks to exchanges between botanists and collectors, its colonization took off.

Around 1884, the plant arrived in Asia, where it was offered hospitality in the botanical garden of Java. How it managed to conquer the entire continent in just a few years is still not clear. Some say it escaped from the botanical garden immediately, taking advantage of the reflux of water following a flood, and that, once it got to a river, it never stopped. Others contend, more romantically, that a Thai princess, having seen the water hyacinth in the botanical garden in Bogor in 1907 and immediately fallen in love with it, brought home some exemplars to her palace pond, from where, having no natural enemies, the plant spread throughout all of Thailand in just four years.

It got to Australia in 1890, as an ornamental plant for ponds. In 1895, it was found living free in New South Wales, and in

1897, botanists at the Royal Botanic Garden in Sydney worried about the speed with which it had colonized all of the garden's ponds and waterways. By the early years of the twentieth century, it had crossed the frontiers of New South Wales and entered Queensland. In 1976, entire fluvial basins were completely covered with the plant.[27]

The water hyacinth arrived in Africa in successive waves starting with the end of the eighteenth century and continuing down to our own time. It was first sighted in Lake Victoria in 1989. By 1995, 90 percent of the Ugandan part of the lake was covered with it.

In its continuous expansion through the tropical areas of the planet, the water hyacinth inevitably found its way to the United States toward the end of the nineteenth century. In this case, too, its passe-partout for entry into the States was the beauty of its efflorescence.

In 1884, during the New Orleans World's Fair, also known as the World Cotton Centennial, a group of Japanese visitors paid homage to local authorities and the organizers of the event by presenting them with some exemplars of the water hyacinth. The gift was greatly appreciated. As usual, the plant's fascination was in the beauty of its flowers. So, with the intention of depriving as few people as possible of the pleasure of enjoying the flowering of the guest from Japan, the exemplars were subdivided among the principal public and private gardens throughout the state that were endowed with opportune aquatic areas. The impact was instantaneous. In a few short years, the nearly supernatural ability of the water hyacinth to propagate itself along waterways rendered the plant ubiquitous in many states of the South.

The spread of the species was so fast and unstoppable that it quickly became a serious problem. As early as 1897, in Florida, the main waterways were found to have as much as twelve pounds of *Eichhornia crassipes* per square foot. No one managed to contain the spectacular spread of the plant, and the breathtaking speed of its dispersal constituted a risk for fish and aquatic animals as well as for numerous water-based economic pursuits. Not the least of these last was navigation. In some rivers, the blanket of vegetation was so thick and extensive that it was impenetrable for boats. The need for a remedy was urgent. But what could be done to block an advance that looked to be unstoppable? Proposals sprouted up everywhere: from the use of natural enemies (but none of the ones proposed seemed to have the slightest impact on the spread of the plant), to projects for machine harvesting carried out by specially modified vessels, all the way to the proposal by the Department of Defense to pour oil on the plants and set them on fire. All legitimate proposals but also totally ineffective.

It's at this point in the story that an extraordinary character enters the scene, the very symbol of the saga of the Wild West and indisputable hero for millions of people in the United States and elsewhere. Allow me to introduce him by using the original words with which a speaker introduced him to a live audience at the beginning of the last century: "I am going to introduce you to a man who knows the cruel edges of war . . . a soldier. A scout whose name has filled both hemispheres with stories of his daring and loyal service . . . the rider of the bad lands between the lines, who trusts his own knowledge some, providence a great deal, and the sound legs and good horse sense of his steed perhaps most of all . . . I am honored, in being permitted to present . . . the only man in America who [knows] the darkest shades of darkest Africa . . . Major Frederick R. Burnham."

His story has the flavor of the incredible; the number of heroic adventures that he participated in over the course of his life is literally incalculable. As is often the case, describing things that happen in the United States requires units of measure bigger than their European counterparts. Grander spaces, more magnificent buildings, more powerful machines, longer trains, and certainly more awe-inspiring heroes. Hundreds of books have been written about Burnham's life. I certainly do not have the space here to trace, even briefly, a plausible portrait of such an outsized personality, but I must ask you to put together two hard and simple facts, based on incontrovertible evidence.

Short in stature (just under five feet four inches tall), Burnham shared that Napoleonic aptitude for command and authority that many men of reduced dimensions tend to develop. His body, though small, was incredibly compact and appeared to be made of some invulnerable fiber. He managed to hold up under deprivations and wounds that would have killed anyone else. It was said that, like a cat, he was possessed of the classic—in Italian terms—"seven lives," the number of lives that one of our heroes would need just to collect—without even pretending to play a starring role in them—the countless adventures that highlight the life of Major Burnham. I will try to run through just the main ones.

He is born to missionary parents on a Dakota Sioux reservation. Still in swaddling clothes, he survives an attack led by Little Crow during the Dakota War of 1862. At age twelve, he is already on his own in California, working as a rider for Western Union. At fourteen, we find him as an expert scout following the tracks of the enemy in the Apache Wars. He takes part in the expedition to capture and kill Geronimo. He fights in the Pleasant Valley War. He learns to shoot with a gun in each hand while riding a horse at a gallop. He becomes deputy sheriff of Arizona's Pinal County. In 1893, concluding that the American frontier has by now become a tame place, he sails with his wife and son for South Africa to join up with the British pioneers in Matabeleland (subsequently better known as Rhodesia). During the thousand-mile hike from Durban to Matabeleland, the war between the British and King Lobengula of Matabeleland breaks out. Burnham enlists and becomes a British national hero. In 1895, he leads a British expedition to Northern Rhodesia. He shares in the discovery of numerous copper mines and is elected to membership in the Royal Geographical Society. In 1896, he participates in the Second Matabele War, during which he meets Robert Baden-Powell and together they draw up the plans for an organization that will see the light a decade or so later: the Boy Scouts. He returns to the United States, where the Spanish-American War is being waged, but he gets there late and the fighting has already come to a close. In 1900, exploring the Klondike, he receives a telegram asking him to serve as head British scout in the Second Boer War, and, without hesitation, he catapults from the Klondike to Cape Town, on the other side of the globe. During the conflict, he spends most of his time behind enemy lines, blowing up bridges and train tracks. Twice he is captured and escapes. He is gravely wounded but survives. He is invited to dinner by Queen Victoria. Her son Edward VII offers him a commission as a major in the British Army. From 1902 to 1904, he leads mining expeditions in

Africa. Then he participates in the First World War. He works in counterespionage. He discovers oil in California, makes a fortune . . . and so on, and on, but I'll stop here.

In short, there should be no doubt, Major Frederick R. Burnham is an American legend. So, in 1910, when together with the senator from Louisiana Robert Broussard and Burnham's former sworn enemy Fritz Joubert Duquesne,[28] he begins a campaign to convince the United States Congress to authorize the importation of hippopotami, nobody claims he is crazy. On the contrary, the idea seems ingenious: import hippos from Africa to be raised along the rivers and swamps of Louisiana, so they will eat the water hyacinth and produce all the meat that the United States in those years so desperately needs. The argument is convincing and not without a certain charm. Before the congressional committee convened to decide on this bizarre request, Burnham asks why Americans stubbornly insist on consuming only cows, pigs, sheep, and chicken. Are they supposedly American animals? No, they were all imported by Europeans centuries ago. So why not import hippos? With time, Burnham tells the committee, a roast hippo will become as natural for Americans as a pork chop or chicken soup. His reasoning is wrinkle free, but by just a single vote the committee fails to usher in the revolution.[29]

I don't know if introducing hippos would have resolved the meat shortage. Maybe. I don't have the competence to imagine what would have happened to hippopotami in an environment so different from their native habitat. I do know, however, that hippos are not domesticated animals. I don't think it would have been so easy to raise them. I have fewer doubts, however, about the fact that their presence wouldn't have done much to contain the spread of the water hyacinth. On numerous other occasions, people have tried to rein in plant species thought to be invasive by introducing potential predators. These efforts have almost never produced the hoped-for results. Often they have created problems worse than those they were meant to resolve. In the most fortunate cases, they succeeded only in introducing a new species to worry about. If that one committee member who cast the decisive no vote had voted yes, there might be hippos today in the United States, but I have no doubt that they would still be swimming in rivers and swamps infested by the water hyacinth.

1 2 3 4

8°f'

Desert

Aspidistra Se...

Poulonia Coast

Yeneu

Imtatuckt zire

a

t.i'

b

Phormium
Cap.

g.t'

Adonide   Alatonia

Betee

Alium

Legorus Islands

Port
Anaxba

h°ö

Antaria

Artilber

Pyrus
Herminui

z.i

4°26'   31°53'   15°12'   12°25'

3

CAPTAINS COURAGEOUS

type species Coconut tree – domain Eukaryota – kingdom Plantae

division Magnoliophyta – class Liliopsida – order Arecales

family Arecaceae – subfamily Arecoideae – tribe Cocoseae

subtribe Butinae – genus *Cocos* – species *Cocos nucifera*

origin Southern India (presumed) – dispersion Tropical areas worldwide

first appearance in Europe Sixteenth century

 Today, just about everybody knows that plants are able to disperse themselves, even in places far away from their land of origin, thanks to their seeds. However, that has not always been the case. All you have to do is go back in time just a little, to the first half of the nineteenth century, to get to an age when no one had any idea how plants had been able to get to the places where they were found. How to account for the hundreds of different species that explorers discovered on desert islands never previously explored? Perhaps God had created different species depending on where they were on Earth? Or maybe there had been multiple creative events, one for each part of the Earth? These were the theories entertained by the majority of scientists.

Keep in mind that, before Darwin, the prevailing idea for explaining the multitude of living species was that they had been created one by one. Each different from the others.

Or else, as some proposed, there had once existed connections linking the emergent continents, across which plants had been able to spread? This would have explained how the flora of an island such as England was not all that different from the flora in nearby regions, on the other side of the English Channel. Let's not forget that the theories of plate tectonics and continental drift were first expounded by Alfred Wegener in 1912, but it was not until the second half of the twentieth century that a long and indisputable series of discoveries convinced a highly skeptical scientific community of their validity.

In any event, neither the creationist theory nor the theory of connections between continents convinced Charles Darwin. He was not totally opposed to the second—his correspondence has some references to his being aware that important changes in coastlines had taken place in the past—but his personal view of the question was different. He was convinced that plants were able to disperse their seeds over even great distances, by utilizing carriers such as wind, animals, and water. Especially water. Darwin didn't see any other possible way to explain the colonization of islands very distant from any other land. Just as humans had reached every corner of the Earth by sailing the seas, so, too, plants must have dispersed themselves throughout the world thanks to water.

Obviously, the considerable difficulties posed by such a theory were not lost on Darwin. For example, there was no evidence at all that seeds had the capacity to survive in saltwater. That they could do so for a few days or at most a few weeks seemed reasonable. But what about staying in the water for the several months needed to reach the most distant lands? To Darwin, too, this eventuality seemed highly unlikely. In any case, there was not much to discuss; before he could decide which theory was right, he had to come up with some evidence to support it. Unlike the creationist theory, or the theory based on the presumed past presence of land bridges between the continents, for which finding evidence was no easy task, the theory of the aquatic dispersal of seeds was rather easy to test. It was not impossible to imagine experiments to verify the capacity of seeds to survive in saltwater. Darwin procured a number of seeds of common species such as oats, broccoli, flax, cabbage, lettuce, onion, and radish and put them in bottles containing a good amount of saltwater. The bottles were then placed in different environmental situations: some in the garden in front of his house, others in the basement, others even in ice water, so as to evaluate the effects of the temperature in different environments. At periodic intervals, a certain number of seeds were extracted from the bottles and planted to evaluate their capacity to germinate.

The results were good but not inspiring. Many species germinated perfectly after a few days of exposure to the saltwater, but when left for longer periods, their percentages of germination were drastically reduced. Some seeds, moreover, in the experimental conditions to which they were subjected, created rather unpleasant effects. In saltwater, cabbage, broccoli, and onion seeds, as Darwin writes in an article published in 1856,[30] produced offensive odors "in quite surprising degree."

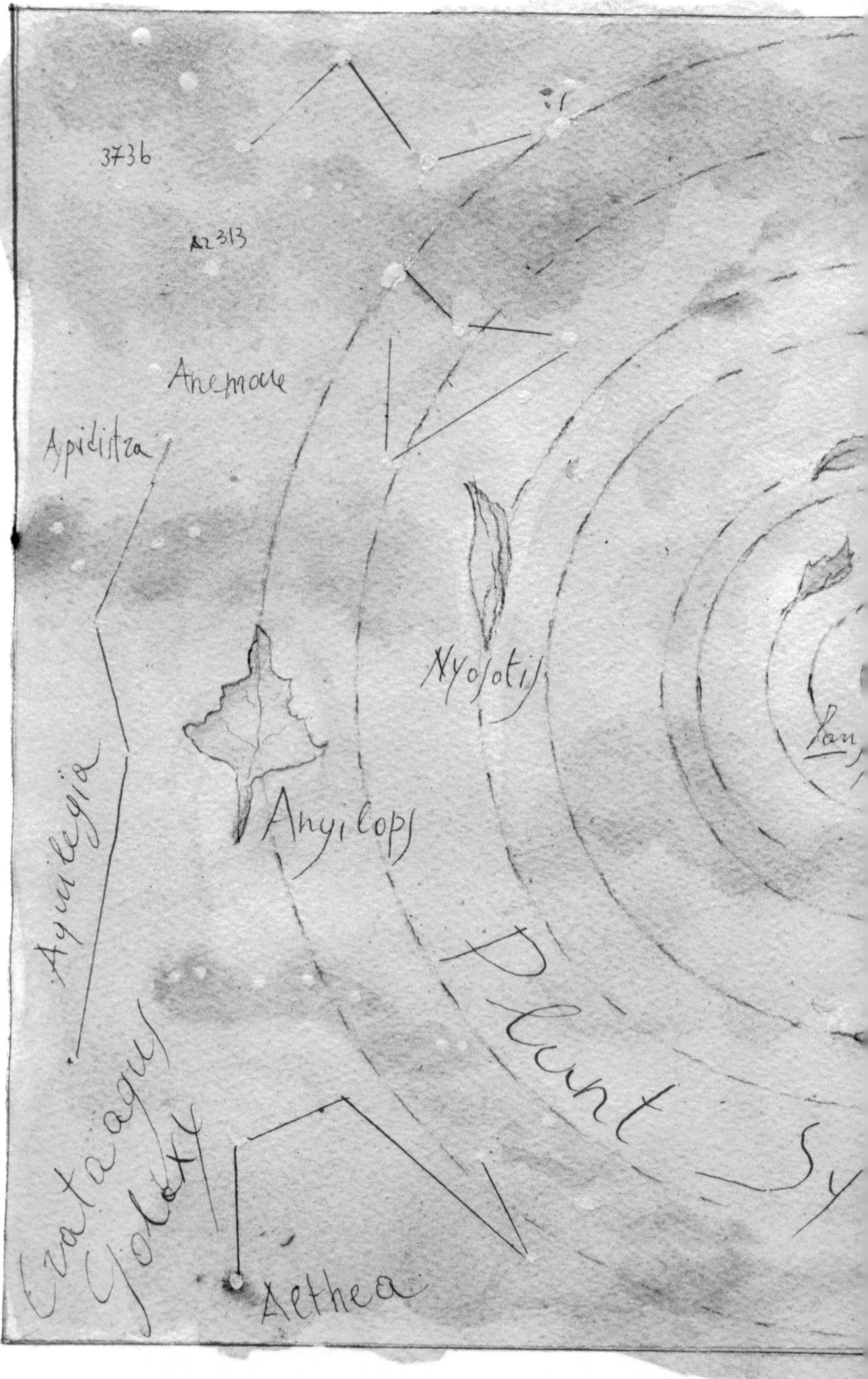
3736
Anemone
Aspidistra
Aquilegia
Angilops
Myosotis
Plant
Crataegus
Galaxy
Althea

Aicca pu Way

z215

Molva jga

Alijlo

Cezis

Cyprypedyum

2713 l

...solis

a

Lethines

...trifile

...ral System

Asphodelus

Nevertheless, "neither the putridity of the water nor the changing temperature had any marked effect on their vitality."

Notwithstanding the bad odors, Darwin was, all in all, quite satisfied with the results obtained and wrote about them enthusiastically to one of his dearest friends, Sir Joseph Dalton Hooker, celebrated botanist and director for twenty years of the Royal Botanic Gardens, Kew. Hooker, however, did not appear to share Darwin's enthusiasm. He saw an enormous flaw in the argument. Seeds, he wrote in his letter of response, usually *do not* float. His friend's simple observation, which he had not considered, threw Darwin into despair. In a letter dated May 15, 1855, he confessed to Hooker that, in light of his friend's note, he feared he had wasted a lot of time "salting the ungrateful rascals for nothing."

Darwin did not limit himself to experimenting with saltwater. He figured that fish might also play a role in the dispersion of seeds. How can it be, he asked himself, that land animals should play such an important role in the dispersal of seeds, and marine animals should not? To verify his hypothesis, he began a series of experiments providing for the ingestion of seeds by fish, but he was decidedly not on a lucky streak. In the same letter in which he complained to Hooker about not having considered that "if seeds sink, then they cannot float," he recounts the misadventures he was undergoing for the sake of those "horrible seeds." In an attempt to verify whether or not fish actually ate seeds, he took himself to the Zoological Society to make some observations. This is his firsthand account: "Everything has been going wrong with me lately; the fish at the Zoological Society ate up lots of soaked seeds, and in imagination, they had in my mind been swallowed, fish and all, by a heron, had been carried a hundred miles, been voided on the banks of some other lake and germinated splendidly, when lo and behold, the fish ejected vehemently, and with disgust equal to my own, ALL the seeds from their mouths."

But it took a lot more than that to stop Charles Darwin. He carried on with his experiments and noticed that, despite everything, some seeds were able to float for a long time. Asparagus seeds, for example, floated for twenty-three days if fresh and eighty-six days if dried—a number of days that would make them hypothetically able, according to his calculations, to travel 2,800 miles, transported by the oceanic currents.

It must be said that Darwin addressed this question of the aquatic dispersal of seeds in a manner atypical of his standard approach. For example, he stubbornly insisted on looking exclusively for evidence in support of his theory through experiments and not by gathering evidence in nature, as he had on many other occasions. Why did he not search directly on the English coasts for the presence of seeds that may have arrived there from distant lands? Why did he not ask his many correspondents throughout the world to inform him about the presence of seeds on beaches? If he had done so, he would have realized immediately the partial validity of his theory. Not all plants, in fact, are capable of dispersing their seeds in saltwater. On the contrary, decidedly few plants succeed in this enterprise. The information available to us today shows that of all the 250,000 known species of flowering plants, only about 250 of them (0.1 percent) produce seeds that can easily be found on beaches. Half of these seeds have the capacity to float in seawater for more than a month and remain vital. This seemingly modest number does not include those species whose seeds are dispersed by remaining attached to parts of plants, branches, rafts of floating vegetation, et cetera. Few species, therefore, are able to swim. Even among plants, great navigators are not all that common. And that's what makes them so interesting.

COCONUT, FRUIT DIVINE

You can't write about the coconut palm tree (*Cocos nucifera*) without mentioning, at least briefly, the use that is made of their fruit. For many of the world's populations, the coconut is the equivalent of what wheat is for Europeans: a basic foodstuff that guarantees survival.

Practically every part of the coconut is used for something; it is like one of those multipurpose Swiss Army knives. In a single compact container we find hypercaloric food and drinking water, fiber to make ropes, and a shell that can be turned into charcoal or, if necessary, a practical floatation device. It is not surprising that in some cultures, especially in Southeast Asia, the coconut has been elevated to the status of a true and proper divinity, a god that many human communities depend on for their survival.[31]

Among the various cults of the coconut, one excels and stands out from the others for such extravagant reasons that I think its story is worth recounting, if only because, unlike all the other cults we know about, this cult was founded in the most unlikely place and time one can imagine: in Germany under Kaiser Wilhelm II at the beginning of the twentieth century.

The story begins in Nuremberg, a city in Bavaria where August Engelhardt was born on November 27, 1875. His is the lead role in our story, as the founder of *Sonnenorder*, the Order of the Sun, a cult of nudist sun worshippers who nourished themselves exclusively on coconuts.

Before starting to work as an assistant pharmacist, August Engelhardt studied chemistry and physics. During this period, he developed his ideas about the necessity of promoting greater contact with nature with the objective of improving people's health. He participated actively in the *Lebensreform* (lifestyle reform) movement, a bunch of *ante litteram* hippies who promoted sexual liberation, alternative medicine, and, in general, a life in contact with nature, based on precepts of vegetarianism and nudism. Engelhardt's ideas, however, were much more radical than those of the *Lebensreform*. Being vegetarian was not enough to ensure a long, healthy, happy life. To achieve that objective, something more extreme was needed. You couldn't take nourishment from all plants. All plants were not equally sacred. Some plants, by their very nature, were close to the sun god, while others were not. Humans should get their nourishment as much as possible from the fruit of the most sacred of plants: the coconut palm tree. Any deviation from this diet would provoke aging, unhappiness, and illness.

It jumps right out at you that a lifestyle like the one championed by Engelhardt, where nudists, devoted to free love, gather and eat coconuts, poses some practical problems for its application in Germany. So, in July 1922, having pocketed a healthy inheritance, August embarked for the Bismarck Archipelago, today part of Papua New Guinea, where he arrived on September 15. Here he bought, for 41,000 marks, a 185-acre coconut and banana plantation on the island of Kabakon, a coralline atoll whose remaining 125 acres were a protected reserve with forty Melanesian residents. The only white man on the island, August built a three-room house, took off his clothes, and started eating tropical fruits. During his stay on the island, he deepened his philosophical relationship with the coconut palm tree, which he had never before laid eyes on, and took from it the following idea: since the sun is the divine life source, and the coconut is the fruit that grows closest to the sun, it follows that the coconut must be the best food for humans to nourish themselves on. He even went so far as to say that if humans were to eat only coconuts they could become immortal. Meanwhile, however, he developed an ulcer on his right leg and saw this as a clear consequence of his having over-indulged in the past on other tropical fruits, deviating from a rigid diet based on the coconut. From then on, for the rest of his life, he would eat nothing but coconuts.

Being the world's only pure coconutarian, however, was not enough. Keeping to himself this knowledge, which could have done so much to improve the fortunes of humanity, made him unhappy. He wanted to spread and enlarge the base of the cult. So he propagandized his idea in Germany and, as a practical incentive, offered to pay ship's passage to any newcomers. The results of this campaign, though modest, did not take long to show up. New proselytes arrived on Kabakon—few, and not all at once, but they arrived, expanding the community up to a maximum of thirty or so members. Many of the new members died a short time after arriving on the island, from malnutrition, infections, or malaria. This was no small problem. The fact that the mortality rate among the coconutarians of Kabakon was much higher than among the Melanesian residents of the same island didn't reflect well on Engelhardt's diet. The German authorities of German New Guinea, practical as always in problem management, required each new arrival to Kabakon to pay a handsome security deposit before receiving an entry visa. The money, they explained, would be used to pay for the hospital care that they would certainly come to need. Later on, when the living conditions for the coconut worshippers became unsustainable, the authorities prohibited any further entries to the island, thus effectively decreeing the end of the community.

Engelhardt was on his own again, his only company, from time to time, coming from some surviving member of the community or some passing German tourist, who inevitably asked to have his picture taken with August. That's how we see him today, in an extended series of snapshots, with his long hair and beard, naked (on the occasion of visits by tourists he covered his genitals with a piece of cloth) and still emaciated and gaunt, with a growing number of bandages on his legs to protect the numerous ulcers caused by malnutrition. On May 6, 1919, his body was found dead on the beach. The last of his followers, laid up in the hospital in the capital of German New Guinea, Kokopo, would die four days later, on May 10. With his death, the epic of the coconut worshippers came to an end.

Now, if you are asking yourselves why in the world I have told you this story, please know that I have done so because people

who are, let's say, eccentric—even if "nutty as a fruitcake" might be more appropriate—like our August are the kind of people I like, and I find their stories fascinating. A good reason in and of itself, but not the only one. A corollary to Engelhardt's activities, in fact, will give me the chance to talk to you about an important question related to the dispersion of the coconut palm tree. But you'll have to have just a little more patience, because the story of August Engelhardt is not over yet. Or, rather, his earthly story, yes, but not his legacy.

To his followers who joined him on the island, just as to the many tourists who asked to meet him over the years, Engelhardt always made a present of a certain number of coconuts, so they would disperse the divinity among the atolls and islands they would visit. Something that many of them, with the fervor typical of neophytes, actually did very passionately, actively contributing to the dispersion of the coconut palm tree. Because even today it is not clear how the coconut palm spread around the world or where it originated.

For centuries, the fruit of the coconut palm tree must have been a mystery to many places in northern Europe. Until the news arrived in Europe that in the distant corners of the world there existed palm trees capable of producing giant fruits the size of a baby's head, when those floating coconuts reached the coasts of Norway and Ireland, they must have had the air of the obscure and inexplicable. What were they and where did they come from, those voluminous round objects that every so often washed up on the beaches? For centuries, Europeans knew little or nothing about the coconut. Marco Polo writes about it briefly in his *Travels*, and later, so does Antonio Pigafetta, in *The First*

Voyage Around the World, his chronicle of the expedition on which Ferdinand Magellan, from 1519 to 1522, completed the first circumnavigation of the globe. But neither is more than a brief mention or a hurried description. Indeed, there was still nothing known about the coconut palm until the Spanish and Portuguese, who had learned to appreciate the fruit's qualities during their explorations of Southeast Asia, began to disseminate it in all the areas of the planet where the climate made its cultivation possible.

Now, even before the Spanish and Portuguese, starting in the 1500s, began a capillary dispersal, the coconut palm tree was already present in very distant places on Earth. How did it manage to reach them? Were humans responsible for its dispersal, in a manner similar to that of the followers of August Engelhardt? Or was it the navigational ability of its seeds that allowed them to cross the oceans? And, above all, is it an original species of the Americas that spread to the Far East, or vice versa? Let's see if we can sort out the controversy.

The first question is: When Columbus arrived in America, was the coconut palm tree already present or not? The answer is by no means certain. None of the first explorers to arrive in America cites the presence of the coconut palm. Christopher Columbus, Amerigo Vespucci, Hernando de Soto, Juan Ponce de León, and their colleagues never make reference to anything that might resemble the coconut palm tree. The only one who comes close to identifying it, in Nicaragua, is the historian Gonzalo Fernández de Oviedo, but some of the characteristics of the fruit he describes would seem to belong to a different species of palm tree. In any event, even if we were sure of the presence of the coconut palm in very small areas of Central America, it would still be a mystery why it did not spread to the rest of Central and South America. Here, in fact, we know with certainty that the coconut arrived only later, thanks to the Portuguese plantations.

The American origin theory seems to be favored by the fact that other Cocoseae[32] are not found in Asia and that the closest relative to the coconut palm grows in South America—a little too little to support the theory. In southern Asia, in fact, the coconut has been known forever, and, furthermore, its highest levels of genetic variability are found there, as normally happens in centers of origin. Taken altogether, as I'm sure you are beginning to understand, it is one big muddle, which, as captivating as it is for oddballs like botanists, I realize may leave other people rather cold. So let's drop it here and get to the point.

A strong thrust in favor of the presumed South American origin of the coconut palm comes from the ideas not of a botanist, but of a famous Norwegian explorer, archaeologist, and anthropologist. Thor Heyerdahl became famous in the 1950s, thanks to his adventures on the boat *Kon-Tiki*. In 1947, using a vessel made from balsa wood, according to Incan tradition, Heyerdahl set sail from Callao, the most important port of Peru. Driven by the Humboldt current, he managed to make it to the Tuamotu Archipelago, in present-day French Polynesia. Thanks to this voyage, Heyerdahl demonstrated the theoretical possibility that South American peoples could have reached and colonized Polynesia in the remote past, bringing along with them plants, such as the sweet potato and the coconut. Heyerdahl's theory was that these Native Americans, using boats analogous to his, had been the first humans to colonize Polynesia. Despite the fascination of Heyerdahl's adventures, genetic testing has demonstrated the exact opposite. Indeed, the mitochondrial DNA of Polynesians shows a greater resemblance to the DNA of inhabitants of Southeast Asia than to that of South Americans.[33] The colonization of South America was begun by inhabitants of Southeast Asia, not the contrary.

Thor Heyerdahl based his theory on a series of presuppositions, some of which were of a typically botanical nature. One of them we know already: the coconut palm is present in both Central America and Asia. Another derives from the distribution of the sweet potato (*Ipomoea batatas*), cultivated in South America at least from 2000 BCE and certainly present in Polynesia as early as 1200 CE. Although it is clearly original to South America, how did it manage to get to Polynesia? Heyerdahl believed it had done so on board boats similar to the *Kon-Tiki*, maneuvered by South American sailors. He was wrong in this case, too. Again, DNA testing has recently allowed us to resolve, once and for all, the question of the dispersion of the sweet potato: it is South American, and it arrived in Polynesia long before humans.[34] Question resolved: another great navigatrix. There remains the question of the coconut. We still do not have the so-called smoking gun, but the great majority of scholars believe that it took the opposite route of the sweet potato, arriving in South America from Southeast Asia.

Whether it got to South America on its own or accompanied by humans, the coconut remains one of the great navigators of the vegetable kingdom. Capable of remaining vital in seawater for more than four months and of using the ocean currents to spread throughout the Pacific, wherever it has arrived, like the sweet potato, it has changed the story of entire continents.

THE CALLIPYGIAN PALM

I'll say this right away, so there won't be any doubts: the *coco de mer*, or *sea coconut* (*Lodoicea maldivica*), that is, the palm tree to which we are about to turn our attention, does not disperse itself at all. On the contrary, we could say that it is one of the least mobile vegetable species in existence, as witnessed by its limited distribution to just the two islands of Praslin and Curieuse in the Seychelles. Nevertheless, as you will see, its story also has something to do with the sea. And if you are marveling at the fact that a species endemic to the Seychelles is called *maldivica* (or "of the Maldives"), then you are not familiar with botanists. By now, you ought to know that they are a strange breed. So strange, in fact, as to change the name of a species from the Seychelles, which had been reasonably named *Lodoicea sechellarum*, to *maldivica*. Agreed, it is rather odd to call *of the Maldives* a species endemic[35] to the *Seychelles*, but that's still something that one can let slide. Something, on the other hand, that cannot be forgiven is that, even before that, in one of the irrepressible impulses that induce botanists, like new Adams, to rename every known plant, they deprived this tree of the name that was most suited for it, *Lodoicea callypige*, the Lodoicea of the beautiful buttocks. And if you had ever seen one of the magnificent seeds of this species, you wouldn't be asking yourself why. Though I do understand that maybe Louis XV of France, to whom the genus name is dedicated, might have a gripe about being associated with such an unroyal name. Nevertheless, the tree in question is really a princess of a palm.

For starters, it holds a certain number of botanical records. It produces the largest wild fruit in nature (ninety-three pounds; some domesticated plants, like pumpkins, can produce heavier fruit), the heaviest seeds (up to thirty-seven pounds for a single seed), the longest cotyledon (up to thirteen feet), and larger female flowers than any other known palm. As if these records were not enough, these giants among seeds have a magnificent shape, which justifies the adjective "callipygian." Indeed, the seed bears an extraordinary resemblance to a female pelvis, on one side and the other.

Up until 1743, the year the French sea captain Lazare Picault, on a mission to map the Seychelles archipelago, saw and summarily described it, the only known parts of this palm tree were its enormous nuts. From time to time, empty ones floated up on the beaches near the Maldives, giving rise to legends about their provenance and their curvaceous features. One of the best-known stories associated the abnormal seed with a mythical tree called Pausengi, which grew somewhere in the ocean south of Java, solidly rooted in the ocean floor. By generating whirlpools around its trunk, this tree inexorably attracted every passing ship incautious enough to get near it. Its crown, on the other hand, was the elected domicile of an enormous bird, perhaps the mythical Roc, which every night, after going out to hunt on land, flew back clutching in its claws immense elephants, tigers, rhinos, and other huge beasts. The fruit of a tree of that kind couldn't help but have superlative properties, and, indeed, it was traditionally believed to be an antidote to poison.

The story of the tree of Pausengi and its astounding fruit appears in an exceptional treatise on tropical botany, *Herbarium Amboinense*, written around the second half of the seventeenth century by the German naturalist Georg Everhard Rumpf (though he preferred to write his name using the Latin orthography, Rumphius). He wrote the treatise during his years on the island of Ambon, in the Maluku Islands archipelago, today in eastern Indonesia.

Rumphius was one of botany's true champions. During his sojourn in the Maluku Islands, he identified and described numerous vegetable species, previously unknown—a huge accomplishment that in Europe earned him the nickname "Plinio Indicus" ("Pliny of the Indies"), and all this despite a series of personal catastrophes. In 1670, at age forty-three, he became blind from glaucoma. In 1674, during an earthquake on Ambon, he lost his beloved wife, Suzanne (whose name he had given to an orchid), and a son. In 1687, a fire destroyed his library and most of his manuscripts and drawings. After years and years in which he managed to reconstruct his lost work, he sent it to Amsterdam to be published, but the ship was attacked and sunk by the French. Fortunately, he had kept a copy, and finally, in 1696, it arrived in Amsterdam. There, the Dutch East India Company decided it contained too much non-divulgeable information, and so it blocked publication of the work for almost fifty years. Rumphius died on Ambon in 1702. His *Herbarium Amboinense* was finally published between 1741 and 1750.

An immense work: seven volumes *in folio*, 1660 printed pages, 695 plates. A marvel. An immense compendium of data, one of my dreams as a bibliophile and an inexhaustible source of delight. Fantastic stories, probable and improbable uses of the various species, legends, imaginary worlds. Rumphius was a typical botanist of an age that is irremediably gone. A world in which poetry and fantasy were still among the tools of the naturalist's trade. When it came to giving names to unknown plants, Rumphius used nothing so banal as *Lodoicea maldivica* or *sechellarum*. Look at these names coined by Rumphius: inkwell root, naked tree, stinking amaranth, adulterer's plant, Saturn's beard, lord of the flies, memory weed, star fish tree, blue clitoris flower, nymph's hair, night tree, scarlet cutlass, sad herb, blind-eyed tree, maiden herb.[36] So, too, when he had to come up with an explanation for those enormous butt-shaped nuts, which sailors collected every now and again during their voyages and whose origin was unknown, Rumphius gathered up the known information, made it available in his treatise, and supplemented what was known with his imagination, deducing that the nuts came from unknown and dangerous lands. In 1743, right during the posthumous publication of Rumphius's masterwork, the unknown land of the callipygian nuts was identified and the mystery of their origin revealed.

But there were a lot of mysteries about this plant, and some of them are still mysteries today. For example, how does it get pollinated? The *Lodoicea maldivica* is a dioecious palm, having, that is, separate male and female plants. When its flowers are in bloom, it is impossible not to notice this. The male trees produce enormous catkins (male inflorescences) with a phallic shape that leaves no room for doubt. Because of their singular and erotic shape, one of the most widespread beliefs on the island was that the trees made love. According to this legend, on dark, moonless nights, the male palms would uproot themselves and move next to the female trees to couple with them. And if, despite all their precautions, someone were to actually see them engaged in these intimate operations, woe betide him! He would die or become blind. Much less poetically, even though there is still no conclusive evidence, it appears that the pollination of these trees is partially anemophilous—the pollen is carried on the wind—and, in part, mediated by a small, brightly colored day gecko, the *Phelsuma*, which frequently visits palm flowers.

Another mystery regarding the *Lodoicea maldivica*, which has only recently been solved, or for which at least a reasonably satisfactory explanation has been found, concerns the enormous dimensions of their fruit and seeds. Why are they so out of scale? The task of a seed should be, as far as possible, to disperse the species, but a thirty-seven-pound seed is surely not the most practical means of spreading oneself around. So, once again, why are these seeds so big and heavy? Whatever the trees' system is for spreading, nothing like these seeds exists throughout the vegetable kingdom. Such a huge investment of energy and material in a single seed is much more reminiscent of the reproductive strategies of some superior animals than it is to those of plants.

Some animals invest a lot in the production of a small progeny to whom they dedicate long and involved parental care. Does something like this happen, perchance, in the plant world? Up until a few years ago, the idea of parental care among plants might have seemed crazy. Even the slightest hint at something of the kind was greeted with disdain. A lunatic hypothesis. Parental care was thought to be the exclusive prerogative of superior animals. It just didn't seem possible to imagine something similar among plants. Then things gradually began to change, and a series of meticulous studies began to demonstrate that care of progeny also existed among plants.

Care for one's own offspring has been observed, for example, in a minuscule cactus (less than 1¼ inches in diameter) called *Mammillaria hernandezii*, originating in a semi-arid zone of Mexico. In the area where it grows, it rains little and intermittently. The plants that live in this habitat are accustomed, therefore, to experiencing frequent cycles of drought. One of the special characteristics of this mini-cactus is that once it has produced its seeds, it has the capacity to conserve them and release them into the environment only when conditions are better for germination. By keeping them inside the mother plant, *Mammillaria hernandezii* teaches its seeds to face the unpredictability of their habitat. Indeed, the seeds experience together with their mother the cycles of rain and drought, learning how to deal with them after they have germinated.[37]

This is clearly an example of care for offspring, but still not exactly parental care. A proper example of the latter is the basis of the solution to another botanical mystery: How do newborn forest plants and trees survive long enough to become autonomous? Forests and woods, in fact, are very dark places, especially at their lower levels. A tree seed that germinates there would not have access to sunlight for a long time. What is the mechanism that allows these newborn trees to grow until they are tall enough to perform photosynthesis? The solution was found just a few years ago. In a wood, most of the trees are connected through an underground network formed by roots and fungi, which live in symbiosis. By way of this network, the adult trees of the clan take care of their little ones, providing them with the sugars necessary to their survival.[38] Parental care, neither more nor less than that found among superior animals, exists among plants and is more widespread than is commonly believed.

Returning to our *coco de mer*, could there also be something similar in this species, as the dimension of their seeds would seem to indicate? In 2015, a brilliant study[39] definitively resolved the enigma, providing a convincing explanation for why the seeds of the *Lodoicae maldivica* are so large.

Let's start from an observation. The environment where this palm tree lives is extremely poor in nutritional resources. Phosphorous and nitrogen, two key elements in plant growth, are present in the soil of islands in very limited quantities. In response to these limitations, the tree has evolved solutions capable of increasing the chances of survival of its own progeny. The solution it has found is astounding and, as far as we know, absolutely *unique* in the vegetable kingdom. To take care of its infants, the sea coconut palm has developed, through its leaves, a system of funnels and gutters to direct water and nutrients to them.

The system works like this: The rainwater that falls on the leaves is directed, by way of these gutters, to the base of the plant. Flowing down from the crown, the water carries with it the residue of nutrient substances—animal dung, pollen, and dead vegetable matter—directing it all to the base of the trunk and fertilizing the soil with phosphates and nitrates. In this way, in the area immediately surrounding the tree, the quantities of phosphorous and nitrogen are decidedly higher. In this situation, the most convenient strategy for ensuring the survival of its progeny is that its seeds fall as close as possible to the mother plant—the exact opposite of what happens with other plants.

The ancestors of the *Lodoicea maldivica* probably used animals to disperse their seeds. Later, when the present-day Seychelles broke off from India, around sixty-five million years ago, the palm found itself without carriers to disperse its seeds. From that moment on, the seeds fell to the ground and stayed there. Consequently, the plants had to adapt to growing in the shade of their parents' crowns. This generated very dense forests formed solely by *coco de mer*, from which the other vegetable species, not adapted to shade, were soon expelled. One consequence of the palm's sedentary adaptation is that, by falling close to the mother plant, the newborn has to compete with its parents and with the other seeds that fall and germinate very close by. In these conditions, the bigger the seed, the greater its energy reserves and, therefore, its chances of survival. There you have it, the solution to the mystery of the mega-seeds: island evolution and parental care. Rumphius would have been satisfied.

Thuia Island

Northern Sea

Tiphia

Terebinthus Glacier

Scilla Channel

15,3° 17,2° 19,1° 20,9° 21,8° 22,7°

4

TIME TRAVELERS

type species Date Palm – domain Eukaryota – kingdom Plantae
division Magnoliophyta – class Liliopsida – sub-class Arecidae
order Arecales – family Arecaceae – subfamily Coryphoideae
tribe Phoeniceae – genus *Phoenix* – species *Phoenix dactilifera*
origin Northern Africa – dispersion Worldwide

first appearance in Europe Around the year 1000

Time travelers exist. At least in the direction that travels from the past toward the present. They are found all over the place. In fact, they are so numerous we don't even notice them anymore. And you know who they are? You got it: plants.

Some species, especially tree species, thanks to their longevity, which is incomparably longer than that of any animal, have traveled through time down to our day from very distant ages. Others, by protecting their embryos inside robust and unalterable seeds, allow their offspring to cross through time and space.

Champions of longevity are common in the plant world. Many species have the capacity to live more than a thousand years.

Some, like *Pinus longaeva*—a name not chosen by chance—boast numerous members capable of living more than four thousand years. Some of them, the true champions, even reach the age of five thousand, or thereabouts. Methuselah, for example, a *Pinus longaeva* (bristlecone pine) that grows in California, has lived an estimated 4,850 years. For decades, this tree enjoyed the special status of the oldest plant in the world. It was the doyen of the *Pinus longaeva* and, therefore, of the entire vegetable kingdom. Its throne began to wobble when it was noticed that many exemplars of this species, despite not having been so fortunate as to have been given their own proper name—an indisputable symbol of privilege in the world of plants—had reached an age close to and, in some cases, superior to that of Methuselah himself.

But that debate has become less interesting since 2008, when Leif Kullman, a professor at Umeå University in Sweden, discovered a red fir tree (*Picea abies*—Norway spruce) that had reached the incredible age of 9,560. If we want to get picky, Old Tjikko—the name Kullman gave to this champion, in memory of his deceased dog—is a tree that has regenerated its trunk many times over the course of its life, rather than being a single tree of great age. Its roots, however, are still the originals, and the act of regenerating the trunk every five hundred to seven hundred years is precisely one of those mechanisms through which plants ensure themselves a longevity nonpareil.

Old Tjikko is indeed the oldest tree in the world. It was here seven thousand to ten thousand years ago when humans invented agriculture, freeing themselves from the necessity of using most of their time searching for food to survive, and it has continued to exist as we have been developing our entire civilization.

Then, of course, there are the great clonal organisms, such as Pando, a 106-acre trembling poplar forest in Utah, comprised of a single genetic individual, which has been propagating itself, unchanged, for more than eighty thousand years. A practically immortal being that has been living since an era so ancient as to be almost incomprehensible to us. To try to get an idea of how long a life Pando has had: eighty thousand years ago the first Neanderthals were appearing in Europe, *Homo erectus* was still not extinct, and another forty thousand years would have to go by before the first exemplars of *Homo sapiens* arrived on the scene.

Even leaving aside these exceptional cases, the average life of many plants is incomparably longer than the average life of animals. It is fascinating to realize that trees, direct and living witnesses of events that, for better or worse, have been for us ephemeral mortals fundamental moments of history, have traveled through the eras to arrive at our own time.

Still there, in Lincolnshire, is the apple tree from which dropped the apple that allowed Isaac Newton to formulate his universal theory of gravity. Still surviving are many of the trees under which Charles Darwin walked to Down House when he was conceiving and writing *On the Origin of Species*. Still living and still growing taller are the oak trees from which hundreds of people were hanged in many states of the United States. Still prospering, at the estate of Les Collettes in Provence, are the olive trees under which Auguste Renoir spent the last years of his life. Even the olive trees in the Garden of Gethsemane, witnesses to Christ's last hours on Earth, are still alive.

The life span of many trees makes them tried-and-true time travelers able to transport—literally—from the past down to us fundamental testimony for the understanding of our history. Through the study of the composition and concentric growth of trees, for example, it has been possible to resolve some of the mysteries of history, such as the sudden retreat of the Golden Horde from Hungary in 1242, when the country seemed almost defenseless in its hands.[40]

Finally, trees have another ace up their sleeves that can be used to send their representatives even into the distant future: seeds. These survival capsules are so perfect in their simplicity as to make those who study them believe that seeds are endowed with supernatural qualities. Seeds are capable of protecting a living embryo in the most difficult conditions imaginable— in water, buried under ice or hot desert sands, at extreme temperatures on one end of the scale and the other, in the presence or absence of air, nutrients, or shelter—for years, for decades, for centuries, in some rare cases for millennia. All this without the embryos that they transport and protect losing their capacity to give life to a new plant as soon as the right conditions present themselves. Seeds are survival capsules that transport vegetable life in time and space, at times succeeding in such grandiose feats that, as with classical heroes, it is necessary to sing their exploits. Here are the stories of three semi-divine champions of time travel.

THE SEEDS OF JAN TEERLINK

Jan Teerlink was a Dutch silk and tea merchant. Son of a pharmacist and grandson of a spice and medicinal herbs broker, he had learned from them as a little boy to appreciate plants as a solid source of good business deals. From his aunt, however, the famous Dutch writer Elizabeth "Betje" Wolff-Bekker, a passionate gardener, he learned to love plants for their beauty and utility. Notwithstanding his indubitable passion for the vegetable world, Jan Teerlink thought of himself as an excellent merchant whose knowledge of plants was only passing fair. He would never have imagined that there would still be talk of him after his death, not for his mercantile skills, but for his botanical sympathies. Much less would he ever have imagined the adventurous circumstances related to wars, colonies, pirates, and forgotten archives through which all this would come about. But let's not get ahead of ourselves.

In 1803, Jan Teerlink, in his capacity as an officer of the Dutch East India Company, undertook a long voyage to Cape Town, South Africa. As any true devotee of plants would have done, having arrived in South Africa, he gave in to the impulse to visit the local botanical garden for an immediate overview of the various species that could be found in nature in that part of the world. The name of the botanical garden, still functioning in the center of Cape Town, is the Company's Garden, and the company referred to in the name is none other than the Dutch East India Company, which established it in 1650 and was still running it at the time of Teerlink's visit.

Initially, like all of the works created by the Company, the garden had to serve a practical purpose. Thus, the Company's Garden started out as a proper farm, capable of producing the fruit and vegetables needed by its ships rounding the cape. Only later did it become a park and, in part, also a sort of botanical garden, or anyway a place to keep a collection of rare or representative exemplars of the local flora. From his 1803 visit to the garden and his discussions with his colleagues from the Company who managed it, Jan Teerlink came away with a good number of plants belonging to species that, for one reason or another, had sparked his interest. Putting into practice what he had learned at home, he correctly placed the seeds of each species in individual paper packets. On each of these, Jan wrote the name of the species or, if it was an unknown species, a more or less detailed description of the plant that had produced the seeds. Some of the packets had the correct scientific or common name of the species; others a summary description of the plant, such as "medium-sized shrub, thorny with small red flowers"; still others a much more original description like "seeds from a tree with crooked thorns," "unknown mimosa," or even "seeds of the wild melons eaten by the savages along the Orange River." In any event, the various prepared and classified packets were carefully conserved in a red leather wallet, ready to face the voyage that was to take them to Holland. Having completed his assignment in South Africa, therefore, Jan Teerlink embarked on the Prussian ship *Henriette* to start the trip back home. Fate, however, had a different plan, and, just a few days away from the Dutch ports, the ship was captured by an English pirate ship. War had just broken out, pitting the United Kingdom against France, which had refused to abandon the Netherlands, lately redubbed the Batavian Republic.[41]

The ship was sequestered, and its cargo of silk and tea became spoils of war for the English pirates. All of the documents, however, including Jan Teerlink's red leather wallet, were sent to the High Court of the Admiralty and shortly thereafter to the Tower of London, where they were abandoned until a few decades ago, when they were finally transferred to the National Archives. And here the wallet would have remained undisturbed for who knows how many centuries more if, during a periodic recataloging, a guest researcher from the Royal Dutch Library named Roelof van Gelder hadn't rescued it from the oblivion to which it had been consigned for the preceding two hundred years.

Through a series of fortunate circumstances, not least the fact that the name Teerlink and the locality where he was born, Vlissingen, both engraved in gold, were known to van Gelder, the wallet was opened again and its contents checked. Scattered among the commercial documents, there appeared the forty packets containing the seeds collected in South Africa. The question then was what was to be done with them, and chance, once again, played a fundamental role. The National Archives, in fact, is located in the suburbs of London, in the delightful locality of Kew, a sacred name to the ears of anyone who has ever even remotely been involved in the study of plants. Here, in fact, between Richmond and Kew, is the home of one of the temples of botany, the earlier-cited Royal Botanic Gardens, Kew.

I have to think that the proximity of the two institutions played a role in van Gelder's decision to have the experts at the nearby botanical garden analyze the packets and the seeds they contained. In any event, the forty packets, containing the seeds of thirty-two different species, ended up in the sagacious hands of the experts at Kew, initially and above all so they could correctly identify the species. Nobody, not even at Kew, expected that the seeds, come to them directly from the period of the Napoleonic

Nandina

Nymphaea

Geraystu

Tropaeolum aty

Lydia

Longtolca Passage

95
83
62

Wholey

Jaymium Cap

Quercus Bay

Norus Alba

Gordenia

Port Fozythya

Dyanthus

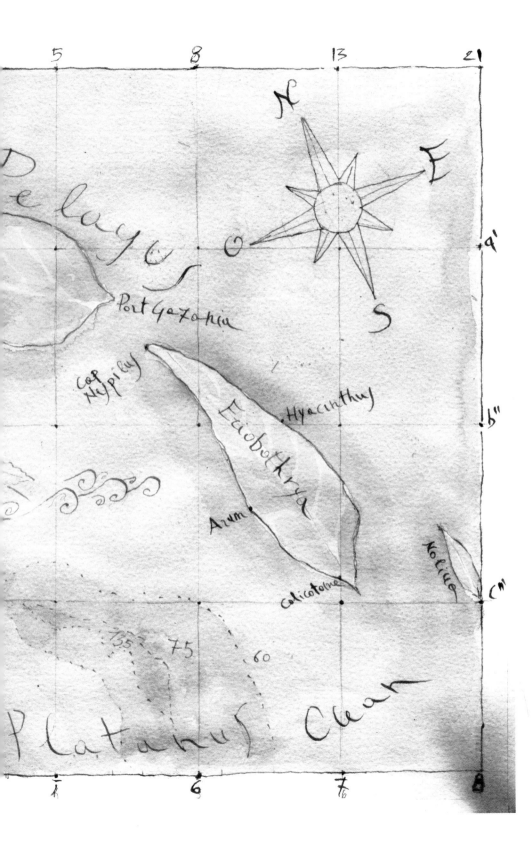

Wars after having been transported on a ship, captured by pirates, forgotten in the Tower of London, and, finally, buried in the depositories of the National Archives, would still be able to germinate. The conditions to which these seeds had been subjected for more than two hundred years were the opposite of what should have been done to preserve their vitality. All the same, it was decided to give it a try, using all possible precautions to ensure the best conditions for their development. To the researchers' amazement, the seeds of three species germinated, two of which survived to give rise to vigorous, healthy plants.

The first species to germinate in force, sixteen seeds out of twenty-five, was a shrub named *Liparia villosa*, whose seedlings, however, did not survive to maturity. Of a group of eight seeds erroneously labeled by Jan Teerlink as *Protea conocarpa*, one germinated, giving rise to a plant later identified as a *Leucospermum conocarpodendron*, which grew and developed perfectly into a healthy adult plant. In 2013, some cuttings from this plant were "repatriated" to South Africa in the magnificent Kirstenbosch National Botanical Garden in Cape Town. The plants that were born from them were called "Princess Elizabeth" in honor of Elizabeth I of England, she, too, a survivor of a period of detention in the notorious Tower of London.

THE DATE PALM OF MASADA

The powerful fortress-palace of Masada rose up, impregnable, atop a pinnacle of brown limestone and dolomite on the border between the Judean desert and the Dead Sea valley, in southeastern Judea, about sixty miles southeast of Jerusalem. This citadel, built around 35 BCE by Herod the Great, who made it his official residence, contained two palaces (one of them three stories high), hot baths, enormous cisterns, fortifications, and a massive uninterrupted sixteen-foot wall around a mile-long perimeter, surveilled by numerous sixty-five-foot towers. This place was the stage for fundamental events in the history of Israel and, as we shall see, for a discovery that has greatly influenced our knowledge of plants.

On the death of Herod in 4 BCE, the fortress passed into the hands of the Romans, who held it until 66 CE. In that year, Masada, after a surprise attack, fell into the hands of a group of rebellious Jews, known as Sicarii,[42] who fought against Rome and anyone who traded with her. They were the most extreme and ferocious fringe group of rebel zealots, and they were known for their violence and the cruelty of their reprisals. Their sadly evocative name is still used today in many languages as a synonym for "killer." Having expunged the fortress, the rebels occupied it and made it their base of operations, the place where they lived with their families and the departure point for their incursions. The Sicarii believed the fortress was impregnable. Additional defensive works, which limited access to a single narrow path overhanging a cliff—the snake path[43]—made even thinking up a plan of attack an arduous task. We can imagine that the Sicarii were so convinced of the solidity of their fortress that they lived there with a feeling of absolute security. The Romans, however, were not used to tolerating open acts of rebellion against the empire. So, in 70 CE, after the fall of

Jerusalem and the destruction of the second temple, Masada remained the only still active center of resistance to the Roman occupation.

The situation could not last long. The moment the rebel zealots captured Masada and barricaded themselves inside it, their fate was sealed. In 73 CE, Lucius Flavius Silva, at the head of the Legio X Fretensis, completely surrounded the base of the rocky pinnacle and built numerous fortified camps to house his troops during a siege that was expected to be long and hard. The disposition of the various camps and the long, uninterrupted wall around the rocky ledge are still traceable in well-preserved and impressive archaeological ruins, which make it clear, even today, that the Romans wanted to make a show of strength against the besieged rebels. These constructions are beyond any reasonable scale, built especially to stagger the imagination.[44] The walls, for example, which Flavius Silva had built even in areas in which, given the lay of the land, they were obviously superfluous (above ravines or crevasses, which nobody could ever have crossed anyway), served only as a warning to the besieged: none of them would manage to escape the wrath of Rome.

The initial plan of forcing the fortress to surrender in the face of the siege did not produce the desired result. Since time was going by and nothing indicated that the zealots had any intention of delivering themselves up, Flavius Silva changed his strategy. The sense of security of the Jewish resistance in the invulnerability of Masada irked him. Rome's military power was based not only on the valor of her armies but also on the extraordinary capabilities of her engineers. In Rome's campaigns of conquest, the construction of roads, bridges, perimeter walls, towers, and aqueducts was common practice. The speed and efficiency with which the Romans, according to accounts of the time, were able to build these works, many of which are still standing and functioning some two hundred years after their construction, border on the miraculous. So, if there were no access routes to the fortress of Masada, the answer was simply to build a new one.

Flavius Silva asked his engineers for a solution that would take his army all the way into the fortress. They came up with one that was simple and ingenious: building an access ramp. Ramp building was part of the technical know-how of the Roman legions. Using ramps that led them up to the heights of besieged walls, along with siege towers and various other kinds of machines to attack defenses, Rome had conquered numerous cities, beginning with Athens. At Masada, however, what was needed was not a normal ramp. The height to be scaled was much greater than any ordinary perimeter wall: at least three hundred feet. That's where the Roman engineers showed just how good they were. Indeed, by using as a base an outcropping of rock to the west of the fortress, they managed to build the ramp very quickly. The conquest of Masada was now just a question of (very little) time. On April 15, 73 AD, the Romans entered the fortress, but all they found was a cemetery. The zealots, in fact, led by Eleazar ben Jair, had all taken their own lives to avoid capture and enslavement. Just two women and two children, hidden in a water pipe, survived to tell the tale.

After the reconquest, the fortress remained in Roman hands until the fifth century. Then it was abandoned and forgotten until, from 1963 to 1965, the Israeli archaeologist Yigael Yadin, helped by thousands of volunteers from all over the world, began a robust excavation campaign aimed at bringing back to light the ruins of the entire fortress-palace and the Roman camps built during the siege. The results of the campaign exceeded even his rosiest predictions, restoring to the world's attention this forgotten fortress and giving back to Israel one of the founding places of its history. Still today, the troops of the Israeli army swear allegiance to the State of Israel inside the fortress with the chant, "Never again shall Masada fall!"

The excavation campaign also brought to light, beyond the great works of masonry, the normal remains of everyday life inside the fortress. Among the myriad rediscovered objects, the ones that are most interesting for our story are probably the most humble and seemingly the least interesting of all: some dates found inside a clay vase, from exactly the time of the fall of Masada. These seeds, cataloged by the archaeologists in 1965, were left abandoned for forty years in a warehouse of Bar-Ilan University in Tel Aviv, and they would be there still, useless and forgotten, if it hadn't been for the intuition of two brilliant Israeli researchers: Sarah Sallon and Elaine Solowey.

At the beginning of our era, all of Palestine was covered by a single and continuous cultivation of date palms (*Phoenix dactylifera*), famous for producing easily dried fruit that maintained a high level of quality even after drying. The dates of Judea were among the most sought-after products of

The Leaf Map

Matthiola

Ophrys

Viburnum Wood

Tropaeolum Village

Nasturtius Sea

Centranthus

Hebe

Vinca

Mattiola Channel

Nigella Beach

Verbascum

Moonia Cap

Nespilous Mountains

Mammilaria City

Mazunta Land

Myoporum Bay

Malva Harbour

Viporina Sands

Convolvus

Port Malobe

the entire area. Apart from their exquisite taste, these dates were known for their presumed antibiotic, aphrodisiac, and medicinal qualities. Well, these palms, so famous in ancient times, have not left so much as a trace. We do not even know exactly when they disappeared, though most accounts indicate they were present, more or less extensively, at least until the year 1100.

Around the fourteenth century, during the reign of the Mamelukes, all the agriculture in the region suffered a terrible crisis. Accounts by European travelers in this period make no mention of date palms. Pierre Belon, who traveled in Judea around 1553, even goes so far as to make fun of the idea that the region could ever have produced the quantity of dates reported in the ancient sources. The causes that led to their disappearance are far from certain. Some scholars lay the blame with the Crusades, others with the Ottoman Empire, but the most likely reason for their decline and, therefore, the subsequent disappearance of this cultivation is to be found in the climatic changes that afflicted the region starting from the year 1000. Around that date, in fact, the climate started to become colder and more humid, reaching a peak around the seventeenth century, followed by a century of strong heat and drought.[45] It is likely that these climatic changes led to changes in the temperature or in the distribution of water and precipitation, damaging irreparably a delicate crop that needs a lot of water and care like the date palm.

Whatever might have been the causes, the fact remains that these cultivations, so renowned in the ancient world, disappeared from the region. It wouldn't be until the 1950s that the cultivation of date palms returned to those places, utilizing modern varieties that had nothing to do with the mythical quality of the ancient ones. It seemed that those ancient palms had disappeared forever until Sarah Sallon and Elaine Solowey—the former a researcher in natural medicine, the latter an expert in the cultivation of date palms and avid hunter of ancient varieties—came up with the crazy hypothesis that one of the seeds collected in the archaeological digs, dating back two thousand years, might be able to germinate, bringing back to life, from that remote time, an exemplar of those palms.

They requested and obtained three seeds from the group found in the fortress of Masada, which in the meantime had been dated from the period between 155 BCE and 64 CE. Elaine Solowey had them hydrated with hot water to activate absorption; then she immersed them in a bath of nutrient and fertilizing substances derived from algae. Finally, on the day of Tu BiShvat, the Jewish feast of the New Year for Trees, which in 2005 fell on January 25, she planted them in sterile soil. Eight weeks later, one of the three seeds had germinated.[46] This was an astounding result. Before then, the oldest seed ever made to germinate had been generated by a lotus plant 1,300 years earlier.[47] If everything went right, an authentic date palm from the golden age of this production would be able to produce again after two thousand years.

There was just one problem left: the sex of the palm. The date palm, in fact, is a dioecious species and so, as we have seen, made up of male and female individuals. If the new plant were female, everything would be fine. On the contrary, if it were male, we wouldn't get to know anything about the quality of those famous dates. There was nothing to do but wait until Methuselah—the name that, in the meantime, had been attributed to the plant—became adult and developed its first flowers. The male name evidently was not propitious: the tree bloomed in March 2012 and turned out to be male—and, like all males, not productive.

Although Methuselah did not give us this ultimate joy, the trail has nonetheless been blazed. Dates are found fairly commonly in archaeological digs, and researchers have already started to attempt germination with other seeds from the same period, conserved in storage rooms in museums and universities. All they need is a little luck. If another seed, this time from a female tree, turns out to be capable of germinating after two thousand years, a new time traveler could make its way to our era to keep Methuselah company, and make us a gift of its delightful dates.

THE SEED THAT CAME IN
FROM THE COLD

All those who have read the masterwork by Varlam Tichonovich Shalamov, *Kolyma Tales*, and I hope with all my heart that they are legion, will have at least two things indelibly impressed in their memories: that Stalinist gulags were an abomination and that everything in Siberia is made of ice. Anyone who has ever read these stories will always associate Kolyma with the idea of freezing cold.

During the years of Stalinism, Kolyma was home to one of the most terrible gulags of the entire Soviet Union: a work camp where living conditions were so frightful that, from the 1930s to the 1950s, nearly a million people lost their lives.[48]

The region of Kolyma, which takes its name from the homonymous river running through it, is one of the coldest places on the planet. Located in the Russian far east, precisely in the northeast corner of Siberia, it is bordered on the north by the eastern Siberian Sea and the Arctic Ocean, and on the south by the Sea of Okhotsk. The average winter temperatures in Kolyma range from −2 to −38°F and can go well below that inland. Here's how, according to Shalamov, the more experienced inmates were able to figure out the temperature: "If there was frosty fog, that meant the temperature outside was forty degrees below zero; if you exhaled easily but in rasping fashion, it was fifty degrees below zero; if there was a rasping and it was difficult to breathe, it was sixty degrees below zero; after sixty degrees below zero, spit froze in mid-air" (translation by John Glad).

Cold is the main characteristic of Kolyma. A cold that kills, but at the same time a cold that halts the decay of organic material. Siberian permafrost, in fact, especially below a certain depth, is a steadily cold environment, with temperatures that remain stable at several degrees below zero for tens of thousands of years. At these temperatures, the hope of finding remains of flora or fauna from the past in conditions suitable to being regenerated from still-vital groups of cells is not a dream.

Permafrost runs through about 8.8 million square miles of the Northern Hemisphere (15 percent of the land on the planet), and in some regions like Kolyma it can reach a depth of several hundred feet. It is not surprising, therefore, that in recent years a growing number of researchers have become devoted to exploring these regions in the hope of finding well-conserved exemplars of extinct fauna. The results have not been long in coming. In 2010, a mammoth cub, named Yuka, cropped up in the permafrost of eastern Siberia, so well conserved that many thought there was a real possibility of bringing it back to life through cloning. In 2015, in the same district of Abyisky, after being buried in the ice for twelve thousand years, two perfectly conserved Eurasian cave lion cubs (*Panthera leo spelaea*) cropped up. Rebaptized Uyan and Dina (from the name of the river Uyandina, near the site of the discovery), they are members of an extinct subspecies of the modern lion. Again in 2015, another Siberian fauna hunter discovered another young offspring, this time a woolly rhinoceros (*Coelodonta antiquitatis*), this one, too, extraordinarily well preserved. All in all, permafrost is turning out to be a gold mine of precious information on many species of extinct animals.

And plants? Although the chances of bringing back to life extinct vegetable species are orders of magnitude greater than those of regenerating animals, the low interest in these living organisms on which life on the planet depends means there are few researchers interested in rediscovering seeds or plants conserved in permafrost. The interest of the public is all in animals. Public interest brings notoriety and funding. Research requires funding. Researchers do research on animals. A simple syllogism that explains why the number of researchers in the world who in various capacities work on plants amounts to a small fraction of the total. Yet, despite it all, how many fundamental discoveries in the history of science have been made by botanists? But I realize I am digressing.

In 2010, a group of researchers from the Russian Academy of Natural Sciences set off from its headquarters in Pushchino, near Moscow, for an excavation campaign in the permafrost along the banks of the Kolyma River. They were searching for animals and plants left trapped in the ice thousands of years ago. They found a promising site for investigation several yards under the surface, in a layer of ice dating back to the Late Pleistocene age. During their investigations, the discovery of a squirrel den, immersed in the permafrost, seemed promising. Dens are always interesting places to go and look around. If you're lucky, you can find some animal that was left trapped. Otherwise, the remains of its daily life—food, excrement, vegetable matter—always provide precious information. In this case, the den had a cache full of seeds and pieces of fruit dating back thirty-nine thousand years.

This discovery was not a novelty. Squirrel dens are often found in permafrost, and their deposits can sometimes contain hundreds of thousands of seeds. This time, however, unlike with many other dens, the seeds seemed, at first glance, perfect. A bizarre idea started to take hold among the discoverers: What if we tried to get some of the seeds to germinate? Obviously, it was clear to all of them that thirty-nine thousand years is an enormous amount of time and that the oldest seed ever germinated until that moment was the two-thousand-year-old seed that gave birth to Methuselah in Israel. It was a fascinating idea, however, and they decided to give it a try. None of the seeds germinated, but under a microscope many showed signs of activation of some tissues or cell groups. Previous research studies had demonstrated, moreover, that, although no seed taken from these dens trapped in the permafrost had ever germinated, many had showed signs of initial growth. A *Rumex* seed had even germinated and grown normally through the cotyledon stage, before it halted and degenerated.[49] In those previous studies, the *Silene stenophylla*, a perennial grass belonging to the Caryophyllaceae family, had always responded rather well. The researchers decided, therefore, to concentrate on this very promising species and to take a different approach. Rather than attempt to germinate a seed from thirty-nine thousand years ago, they tried to regenerate the entire plant from some placental tissue.

The result was extraordinary. The researchers succeeded in growing a perfectly healthy seedling of *Silene stenophylla*, able to develop and produce seeds. What other researchers have dreamed of doing with mammoths, woolly rhinos, and cave lions has been done with a plant from the same era. If any animal from thirty-nine thousand years ago had been regenerated, all the media of the world would have talked about it for weeks. The return to life of a little, insignificant *Silene stenophylla*, on the other hand, was of interest to no one other than a few experts in the field. Yet what marvelous possibilities have been opened up by this research! Squirrel dens full of seeds have been discovered in Late Pleistocene ice not only in eastern Siberia but also in Alaska, the Yukon, and the whole Beringia area.[50] The permafrost is full of frozen seeds and fruits of vegetable species waiting to be regenerated. Countless extinct species could be present in the permafrost, with their inestimable genetic patrimony. Bringing them back to life depends only on us.

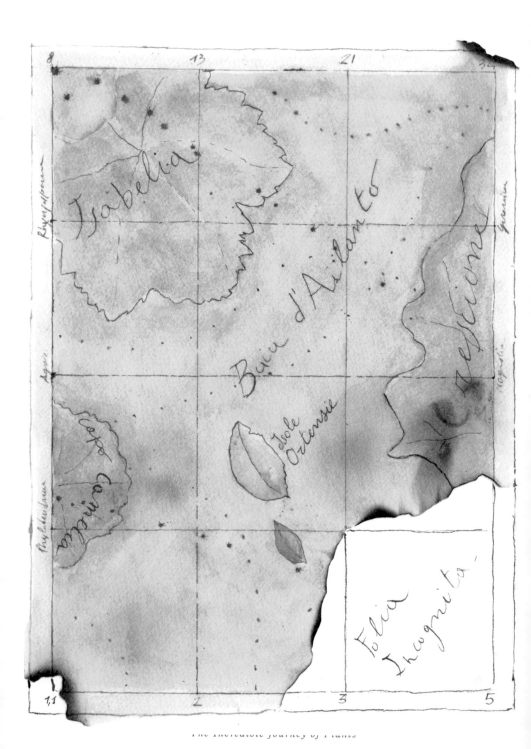

5

SOLITARY TREES

type species Sitka spruce – domain Eukaryota – kingdom Plantae

division Pinophyta – class Pinopsida – order Pinales

family Pinaceae – genus *Picea* – species *Picea sitchensis*

origin West Coast of North America

first appearance in Europe Nineteenth century

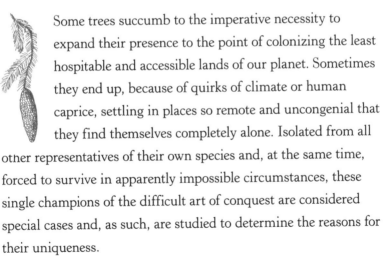

Some trees succumb to the imperative necessity to expand their presence to the point of colonizing the least hospitable and accessible lands of our planet. Sometimes they end up, because of quirks of climate or human caprice, settling in places so remote and uncongenial that they find themselves completely alone. Isolated from all other representatives of their own species and, at the same time, forced to survive in apparently impossible circumstances, these single champions of the difficult art of conquest are considered special cases and, as such, are studied to determine the reasons for their uniqueness.

Although the solitary tree can also be seen as a literary *topos*—symbol of a person who, in spite of everything, resists, indomitable against the arrows of adversity—like so many other of our cultural stereotypes, this one is based on presuppositions that are totally mistaken. Indeed, if you think about it carefully, a solitary tree should not exist. It is a contradiction in terms. Every solitary living being is in some way a contradiction in terms. For there to be life, there must be community with other living beings and, obviously, with other individuals of one's own species. One of the most tragic fates for any living being is to end up as one of just a few individuals—in some cases utterly alone—on the verge of extinction and with no chance to reproduce the species that it represents.

Solitary trees, perhaps precisely because of their irreconcilable nature, have always fascinated the imaginations of artists, especially painters. One of the best-known paintings of this genre is *The Lonely Tree (Der einsame Baum)*, by Caspar David Friedrich.

Why this very famous painting is known by this title is beyond me. Painted in 1822 and conserved today in the Alte Nationalgalerie in Berlin, it features in the foreground a tree, probably an oak, that is anything but solitary or isolated. When it comes right down to it, there is nothing solitary in this painting, unless it's the shepherd leaning against the base of the tree. Some thirty or forty feet away from the starring oak, Friedrich depicts lots of others, and there is even a glimpse of a lovely wood in the background. To be sure, the tree in the foreground is clearly in terrible shape: its trunk is split and crooked; its vegetation shoots up from its base, as happens after severe damage; some of its branches are dead . . . all in all, an oak that has been sorely tested by adversity but nonetheless bravely resists. But at least it is not alone. That much Friedrich allows it. Although his pictorial production certainly does not lack melancholy subjects, Friedrich seems to know that a truly solitary tree is something much more sad and rare than what could be found in the pleasant countryside of Riesengebirge (the mountainous region between Silesia and Bohemia shown in the painting).

There are some truly solitary trees, but not very many. When you happen to come across one, it is always interesting to try to figure out how it got that way. Indeed, every solitary tree, or better, any plant that is isolated, far away from its species mates, almost always has an interesting story to tell. It is not easy to find oneself in this unenviable situation. In impervious places, hundreds of miles from others of their own kind and in climates ill suited to life, these trees survive for unimaginably long periods, testifying to their inexhaustible capacity to adapt to the most extreme conditions. Of the few truly solitary trees that we know about, three of them must be remembered, at least for the places where they live, for the legends that surround them, and, finally, for the contribution they have made to the advancement of our scientific knowledge: the solitary pine tree of Campbell Island, the Tree of Ténéré, and the Tree of Life in Bahrain.

THE SPRUCE TREE OF
CAMPBELL ISLAND

Campbell Island (Motu Ihupuku in Maori) is one of the most remote places on Earth. With a surface area just a little smaller than Nantucket, it is located about 375 miles south of New

Zealand, in the heart of the Subantarctic. It is so far removed from the usual sea routes that it remained unknown to the world until 1810, the year Captain Frederick Hasselborough, in command of the brigantine *Perseverence*, discovered it during a series of expeditions in the Antarctic regions of New Zealand, promoted and financed by the Australian ship owner Robert Campbell (hence the name of the island). The island, incidentally, did not bring good luck to the captain, who drowned there a few months after discovering it, on November 4, 1810.

Campbell Island was then, as it is today, completely uninhabited. At those latitudes, one certainly cannot expect a mild climate. The sun shines on the island very rarely, on average 650 hours per year (in New York, or Rome, there are more than two thousand hours of sunshine per year), and for more than seven months out of the year there is less than one hour of sunlight each day. The temperature hovers around 45°F, it rains constantly, and there are more than one hundred days a year when the wind is stronger than sixty miles per hour. The horrible climate is one of the reasons why no human community has ever settled on this little island in the more than two hundred years since its discovery, except for temporary excursions of hunters devoted to the extermination of the local community of seals (a task promptly completed, as there have been no seals there since 1815) and, more recently, scientists engaged in studying the climate and meteorology of the Antarctic regions.

Campbell Island is not a happy place to live for plants and animals. With that kind of climate, it is reasonable to expect that even trees have no chance to survive. Indeed, the vegetation on Campbell Island is typical of the tundra: mosses and lichens, grasses, a few shrubs, and no trees—with one small but important exception: a magnificent exemplar of *Picea sitchensis*, majestic and solitary dominator of the island's flora. This tree is so far away from every other plant of its species as to be declared officially by *The Guinness Book of World Records* "the loneliest tree in the world."

But how did an exemplar of *Picea sitchensis* ever get to Campbell Island, more than 120 miles from its nearest species mate, which grows on the Auckland Islands? The responsibility for this extreme botanic localization seems to lie with a certain Uchter John Mark Knox, the fifth Earl of Ranfurly, governor of New Zealand from 1897 to 1904. Lord Ranfurly took his duties as governor very seriously. So at the start of the twentieth century, he embarked on a voyage of exploration of the British dominions in his region. The voyage touched every single little island of the Crown, including Campbell Island, which must not have had a great impact on the governor, since the main impression he took from it was that of a completely unproductive, and thus useless, island.

The existence of a territory, no matter how insignificant, that made not even a small contribution to the magnificent destiny of the empire must have seemed to him an intolerable affront, which called for an immediate remedy. So with one of those offhanded bright ideas that often characterize the activities of governors, in all times and latitudes, Lord Ranfurly decided that the island would become a producer of lumber. He ordered that the inhospitable little island be planted with enough trees to cover it with a luxuriant forest that would furnish the lumber needed for building a whole lot of ships. Even the minuscule Campbell Island would have the honor of helping to keep Britannia the ruler of the waves. The small detail that not a single tree grew naturally on the island didn't disturb the governor one bit. Without a doubt, the efficiency of British technicians would resolve this little problem. Campbell Island was to become an austral forest. It was decided.

As often happens, the initial enthusiasm of this grandiose announcement was soon deflated by the lack of any practical action aimed at transforming the island into the forest imagined by the governor. I can almost see those ill-starred experts of forestation on the bridge of the ship, nodding earnestly, taking note of his desires, as the governor indicates with his elegant cane the places he deems most promising for the planting of new trees and suggests the most suitable species. "Down there, those hills declining to the east look to be ideal for fir trees. The lower-lying area to the west, on the other hand, will be perfect for *Picea sitchensis.*"

Nature, as she tends to do, decided differently. Despite the proverbial efficiency of the British Empire and the planting of several hundred trees, just a few years later there were none left. All swept away, frozen, withered by the impetuous freezing-cold winds that blow in from the Antarctic. All but one: our indomitable *Picea sitchensis.*

By virtue of a position partially protected from the elements, or just because it was more robust and adaptable than its companions, our tree held up against the winds, the cold, the lack of sunlight, the cutting of its branches to be used as Christmas trees by the climatologists who happened to be spending the holidays on the island, and against all the other environmental and human abuses. It grew strong and secure despite everything and everyone. From 1902, the estimated year of its birth, to today, our *Picea sitchensis* has continued to grow in one of the most isolated places in the world. This last detail is not incidental. By virtue of its isolated growth, in fact, the lone tree on Campbell Island is also a *one of a kind* for scientific research. Just think; thanks to the results of the studies conducted on this single plant, the suggestion arose to establish the year 1965 as the beginning of a new geological age: the Anthropocene. But let's not get ahead of ourselves.

The scale of geological time is a system used by the international scientific community to divide up the time that has passed since the creation of the Earth. Many readers will have heard the names of some of these periods. The Jurassic and the Cretaceous, for example, are commonly mentioned outside of strictly scientific contexts, while other periods, such as the Ordovician or the Silurian, are much more obscure and known only to specialists. In any event, from the birth of the Earth to today, each moment of its life is described by precise geochronological units: eons (billions of years), eras (hundreds of millions of years), periods (tens of millions of years), epochs (millions of years), and ages (thousands of years). So today, just to give you an idea, we are living in the Phanerozoic Eon, the Cenozoic Era, the Quaternary Period, and the Holocene Epoch. It is a sort of home address that allows us to give a precise order to life on

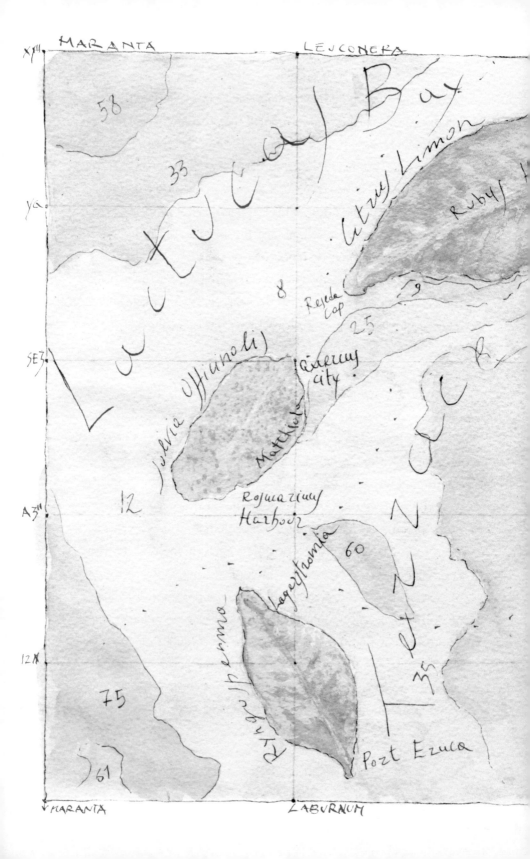

4

Rumex beach

35

67

Cap Laguhania

16

Rouceae

10

Lanitana

Aquifolium

Indica
Hohovz

12

51

157
215
325
502

Ligystio Trench

Robinia

Acuspidata

APOCYGCEOR

SUAVOLENS

HANDEVILLA

our planet in relation to particularly significant moments. This system is very similar to what we all do when we tend to classify our lives in relation to special events (before marriage, after retirement, in the last year of high school, et cetera). The problem, then, is to try to understand (1) whether or not a particular event had such an important impact on the history of the Earth as to define a line of demarcation of a period, epoch, et cetera, and (2) if the event left a trace of itself that is detectable everywhere on the planet.

Some turning points on the temporal scale are determined by great events, such as mass extinctions, and they leave no room for debate. Take, for example, the transition between the Cretaceous Period and the Paleogene Period. In 1980, Luis and Walter Alvarez, father and son, physicist and Nobel Prize winner the former, paleontologist the latter, published their theory on the extinction of dinosaurs caused by the impact of an asteroid.[51] The idea was based on the discovery, made the year before by Walter Alvarez, of a thin layer of clay in the Bottaccione Gorge near Gubbio in central Italy. The clay was datable to the end of the Cretaceous Period and very rich in iridium, an element rare on Earth but rather common in rocks originating in outer space. After that initial discovery, this layer of dust was found all over the planet. The asteroid that struck the Earth sixty-six million years ago thus left an indelible mark in the terrestrial stratification, making it a textbook case of a transition between geological periods. In other cases, the line of demarcation between one period and another is more indistinct and traceable not to a single event but to a plurality of contributory causes. In these cases, the identification of the exact demarcation is not all that simple. Nevertheless, before a geological epoch transition is accepted by the scientific community, it must first be formally recognized by a specially designated international body: the International Commission on Stratigraphy. Okay, we're almost there; just a few more words on what the Anthropocene is and we can go back to our solitary tree.

The term "Anthropocene" (from *Anthropos*, "human"), originally coined by the American biologist Eugene Stoermer, owes its notoriety to the Dutch Nobel Prize winner for chemistry Paul Crutzen.[52] According to Crutzen's definition, the current geological epoch is characterized by human activity, which is very rapidly modifying all environmental characteristics, from soil, to climate, to the presence and distribution of other forms of life. In reality, the idea that humans are actively contributing to environmental change was raised a great deal prior to Stoermer and Crutzen. The first to talk about it, in 1873, was Antonio Stoppani, a priest and Italian patriot known as the father of Italian geology. On identifying human activity as a true and proper geophysical force, Stoppani proposed the idea of calling our epoch the Anthropozoic. Later, the same idea was taken up and broadened by the Russian geochemist Vladimir Ivanovich Vernadsky and later still by Pierre Teilhard de Chardin, the French Jesuit and paleontologist.

Although for 11,700 years, following the end of the last glaciation (the Würm glaciation), the Earth has been officially in the Holocene Epoch, the majority of scientists are convinced by now (beyond all doubt) that human activity has irrevocably altered the terrestrial environment and that the current geological epoch must, consequently, be called the Anthropocene. The evidence is everywhere. One cannot help but notice it. Take, for example, the study published in 2015 by a group of researchers led by Professor Will Steffen[53] on the modification of twenty-four global indicators starting in the 1950s. Twelve of these indicators involve human activities (energy consumption, water consumption, economic growth, population, transportation, telecommunications, and so on), while twelve other parameters, such as biodiversity, deforestation, and the carbon cycle, directly involve the planetary environment. The results are unequivocal. From the postwar years to today the use of fertilizers has increased by 800 percent, the amount of energy consumed has increased by 500 percent, and the urban population has grown by 700 percent.

The impact of these activities, often directly tied to the economic system (some have proposed to call our age the Capitalocene, rather than the Anthropocene), has provoked a series of grave consequences: a worrisome acceleration in the rate of species extinction (so much so that the current period has been called the sixth mass extinction)[54] and the consequent loss of biodiversity, climate change, the exponential growth in the rate of pollution, et cetera. There is no doubt that human activity is changing the planet, unfortunately for the worse.

When did human activity start to turn into a geophysical force? Here the question begins to get delicate. There are at least four different positions on this: (1) With the onset of agriculture, ten thousand years ago. Agricultural activity needs deforested lands to cultivate. Furthermore, once agriculture began, no longer having to employ most of their time searching for nutriment, humans were able to increase in number and initiate the technological progress that would inevitably lead to the way things are today. (2) In the sixteenth century, with the beginning of the great voyages of exploration, the discovery of the Americas, and the consequent redistribution of plants, animals, goods, and diseases.[55] (3) In the second half of the eighteenth century, with the Industrial Revolution and the increase in CO_2 emissions.[56] (4) After the Second World War, with the beginning of the Atomic Age.

Each of these hypotheses has good reasons to support it. In any event, the key to answering the question is to find a global signature, something similar to that global layer of iridium, which left an indelible sign, sixty-six million years ago, marking the end of the Cretaceous and the beginning of the Paleogene. Obtaining this synchronous global trace, containing physical, chemical, and paleontological information able to confirm the contemporaneous jump from one epoch to the next, is not easy.

This is where our solitary tree on Campbell Island comes in, rightfully reclaiming for itself the starring role in this story. Indeed, with the publication in February 2018 of an important scientific work,[57] our *Picea sitchensis* became the famous missing proof. By analyzing the amount of carbon-14 present in the concentric rings produced each year by the tree, the researchers found a peak of carbon isotopes probably produced by the nuclear tests executed in the Northern Hemisphere between 1950 and 1960. Specifically, the peak presence of carbon-14 was found to have been in the closing months of 1965. That this peak was found in the wood of a tree living in a completely uncontaminated place as far as possible from the original source of those carbon isotopes is the unequivocal indication that human intervention in the environment is a global phenomenon. Furthermore, radiocarbon is conserved for fifty thousand years, thus guaranteeing that even tens of thousands of years from now the scientists of the future will still be able to find it. So thanks to a solitary tree that resolutely insisted on growing right there where it should not have, we may finally have the proof, the synchronous global signature, that could be used to mark the start of the Anthropocene.

THE ACACIA OF TÉNÉRÉ

The *Picea stichensis* of Campbell Island has not always been "the loneliest tree in the world." Until 1973, that rather unenviable title rightfully belonged to another exceptional champion of the art of survival in extreme environments: the acacia in the desert of Ténéré. In the middle of one of the most arid places in the world, distinguished by its absolute lack of vegetation, this acacia, towering above the uniform expanse of sand, the only existing tree within a radius of hundreds of miles, was for more than three centuries a reference point for the Azalai, the long caravans of camels with which the Tuareg people transported rock salt from Mali to the Mediterranean. The exceptional nature of this plant, once again, lies in its distance from any other tree (in this case, its distance from *any other plant*) and in its capacity to survive in one of the most hostile places on the planet.

In northern Niger, the climatic conditions of Ténéré are as extreme as they get on planet Earth. To find more difficult conditions you would have to visit other planets in our solar system. Let's start with the name, whose meaning is already evocative enough. Indeed, *ténéré* is the Tuareg word for "desert." Put that together with its location in the north-central part of the Sahara, whose name means, in turn, "desert," this time in Arabic, and the name itself of this desolate region of the world reveals something about its essence. The Ténéré is a desert inside a desert. A searing hot nightmare, classified as a hyper-arid zone, with maximum temperatures that frequently exceed 122°F and one of the lowest amounts of rainfall on Earth, between 0.4 and 0.6 inches per year. This means that several years may go by without a single drop of rain. As if that weren't enough, water is extremely difficult to find underground, too, and the few available wells are hundreds of miles apart. Contrary to Campbell Island, this place has the most hours of annual sunlight on the planet (more than four thousand), and according to a study by NASA the single most sunlit place on Earth is a ruined fort in Agadem, in the southeast of the Ténéré. No vegetation can survive in these conditions. The Ténéré is the classic desert out of *Lawrence*

No way Horizon   Wii

cap Scilla   Extreme cap

The
Unknown
Country

Finis
Terre

Ice Sea

Nowhere Island

Desolate Land

Extremely windy Mountains

...tains

HOJSTLE

Nothing

for

Lonely tree promontory

No Harbour

No Leaves Gulf

of Arabia: hundreds of miles of sand dunes and nothing else. Try to get a look at some satellite pictures to get a better idea of it. How a tree could grow in conditions as inhospitable as these is truly a mystery. The acacia of Ténéré, more precisely an exemplar of *Acacia tortilis*, was so isolated as to be the only tree marked on maps of the region at a scale of 1:4,000,000.

It is believed that this acacia was the last exemplar of a small population of acacia surviving from a time, not so long ago (six thousand years), in which water, not having completely disappeared, could still sustain some forms of vegetable life. European exploration of the Ténéré is quite recent. The first Europeans to reach it were the members of a British expedition led by J. Richardson in 1850. In 1876, a German, Erwin von Bary, followed pretty much the same route as Richardson, and then nothing more until 1906 and the French occupation of the city of Bilma. The following year, a column of 2,500 Méharists[58] (a cavalry corps on camels) managed to cross the Ténéré, following the traditional route of the Azalai. They reached the tree of the Ténéré and etched on its trunk the date of their passage: October 13, 1907.

The tree is mentioned often in the accounts of these first explorations, so often that in the 1930s it was included in European military maps of the region as an important reference point—a sort of lighthouse in the desert, essential for orientation in that otherwise desolate expanse of sand. In a report from 1924, the tree was said to be almost completely covered in sand; other reports recount how the continual movements of the dunes forced the tree into long periods of nearly complete immersion in the sand.

We are indebted to Commander Michel Lesourd of the Service Central des Affaires Sahariennes for one of the first written accounts of the tree:

> *From Agadez, going to the post at Bilma, our automobile convoy arrives at 14:30 on May 21, 1939, at the Tree of Ténéré . . . One must see the Tree to believe its existence. What is its secret? How can it still be living in spite of the multitudes of camels, which trample at its sides? How at each Azalai does not a lost camel eat its leaves and thorns? Why don't the numerous Tuareg leading the salt caravans cut its branches to make fires to brew their tea? The only answer is that the tree is taboo and considered as such by the caravanners. There is a kind of superstition, a tribal order, which is always respected. Each year the Azalai gather round the Tree before facing the crossing of the Ténéré. The Acacia has become a living lighthouse; it is the first or the last landmark for the Azalai leaving Agadez for Bilma, or returning.*

Lesourd's report finally gives us some information on the extraordinary characteristics that made this acacia legendary. Digging a well near the tree, in the hope of finding the same water that the tree drank from, the French were blocked by a layer of granite almost 100 feet thick, which, for their part, the roots of the acacia had easily penetrated. Later, roots from the tree were found more than 150 feet deep.

In 1959, the tree's state of health was no longer so positive. Henri Lhote, member of a geographical mission, wrote in his book *L'epopée du Ténéré*:

Before, this tree was green and flowering; now it is a colorless thorn tree and naked. I cannot recognize it—it had two very distinct trunks. Now there is only one . . . What has happened to this unhappy tree? A truck going to Bilma simply ran into it . . . but it had more than enough space to avoid it . . . the taboo, sacred tree, the one which no nomad here would have dared to have hurt with his hand . . . this tree has been the victim of a machine.

The tree's being hit by a truck was a sad omen of its most unlikely end. Think about it: What are the chances of being hit by a truck in the middle of the nothingness of a desert like the Ténéré? Practically none. A tree could stay there for billions of years stuck in the middle of that desert and never be hit by any mechanical means. How many trees do you know that have been run into by a car on city streets, despite millions of cars passing by them over the course of decades? Very, very few. Now try to calculate the chances that the same tree, the only tree present in a desert for hundreds of miles around, might be hit *twice* by a truck in fifteen years. I am not an ace at calculating probabilities, but I am sure that the chances of winning the first prize in the national lottery ten years in a row are much higher. Yet this is exactly what happened to our acacia in the Ténéré. On November 8, 1973, a Libyan drunk driver managed to materialize this one chance in who knows how many million bazillions and, smacking his truck right into the only tree in the middle of nothing, decreed its end. Maybe it was not really the loneliest tree in the world, but the most unfortunate, yes. No doubt about it.

THE TREE OF LIFE IN BAHRAIN

The Tree of Life (*Shajarat al-Hayat*) in Bahrain is the last solitary tree in this little collection.

Bahrain is a minuscule archipelago in the Persian Gulf, located between Saudi Arabia and the peninsula of Qatar. This ancient tree, one of the most mysterious and fascinating of all known trees, thirty-two feet high, grows majestically atop a sandy hill, completely isolated in the middle of the desert area on the main island of Bahrain.

Although this plant has been known for centuries and there are a myriad of legends about it, from a scientific point of view very little is known about it. The tree, whose name derives from the popular belief that it is the original tree of life recounted in the Book of Genesis[59]—not to be confused with the more famous and fraught-with-consequences-for-humanity tree of the knowledge of good and evil—has been described as belonging to many different species. Unfortunately, there not being any scientific publications dedicated to it, our knowledge of this tree, which, conversely, would have so much to teach us, is little and often confused.

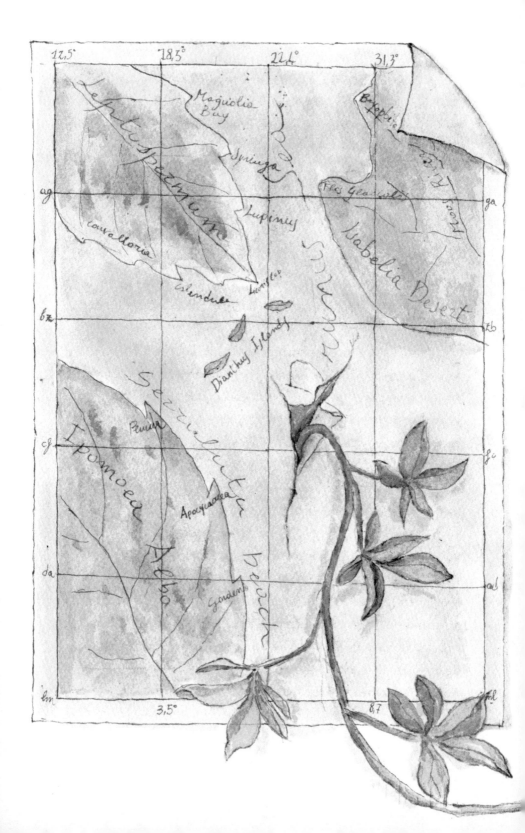

Until rather recently, any attempt to do research on the Bahrain Tree of Life came up against a phantom study conducted in collaboration with the Smithsonian Institution in Washington, DC, which estimated the age of the plant to be about five hundred years. A few months ago, not having succeeded in finding any solid evidence of any publication by the Smithsonian about this tree, I decided to ask for enlightenment directly from the institution. No luck. The person I contacted responded very kindly that she was unable to find any citations of the tree among the studies conducted by the Smithsonian. A situation of uncertainty, therefore, and with no authoritative sources to rely on if not those provided by the government of Bahrain itself, which, some years ago, having intuited the potential that this tree might have, above all for tourism, undertook a reliable series of analyses.

The results of these studies are just as fascinating as the legends surrounding the tree. First, its age: the tree would seem to have survived in a total desert since the second half of the sixteenth century, which would make it by far the oldest of all the solitary trees in the world and, therefore, the one that has succeeded best in adapting to the adverse conditions of its environment. Second, its species: today we know with certainty that the Bahrain Tree of Life is a *Prosopis juliflora*, a tree whose origins are in Mexico and South America, typical of hot, dry, and salty places where few other species are able to survive. Thanks to its taproot, which can reach down to incredible depths;[60] to its small, compound leaves that efficiently disseminate excessive heat and limit water loss; to its capacity to fix nitrogen by virtue of its symbiotic relationship with nitrogen-fixing bacteria; and, finally, thanks to its intrinsic capacity to withstand water with high concentrations of saline—the only water that its roots can find at such great depths in the desert soil—this tree was built to survive in the most difficult conditions imaginable for a plant.

There's more. Not even a champion of extreme climates like the *Prosopis juliflora* could survive in the middle of the desert for five centuries without pulling off some other tricks. In 2010, the government of Bahrain began a campaign of archaeological excavations in the area immediately overlooking the tree, discovering a village that was probably active until the middle of the seventeenth century with a well very close to the base of the tree. This indicated that the tree had been planted there purposely and that, through the centuries, even after the final abandonment of the village, it had succeeded in following the aquifer with its deep roots. There lay the explanation of where the water came from that had assured its survival.

There remained just one small but fascinating mystery: How had a species originating in the Americas arrived smack-dab in the middle of the desert in Bahrain, on the other side of the world, just a few decades after the discovery of America? The most likely route seems to be by way of the Portuguese, who conquered the islands in 1521 and remained there until 1602, the year the archipelago came under Persian dominion. During these years of Portuguese rule, some *Prosopis juliflora* trees must have arrived in Bahrain thanks to the intuition of some Portuguese botanist who thought they had a good chance to survive in an environment so much like that of their original home. The Tree of Life is the only survivor of that first nucleus of trees.[61]

Whatever may have been the adventures that brought the Tree of Life all the way to Bahrain, there remains the extraordinary enterprise of this single plant, which, displaced from the faraway Americas, has managed to grow and prosper, conserving itself for half a millennium in a hostile environment, a living emblem of plants' capacity for adaptation and their ability to resolve brilliantly even the most arduous problems related to survival.

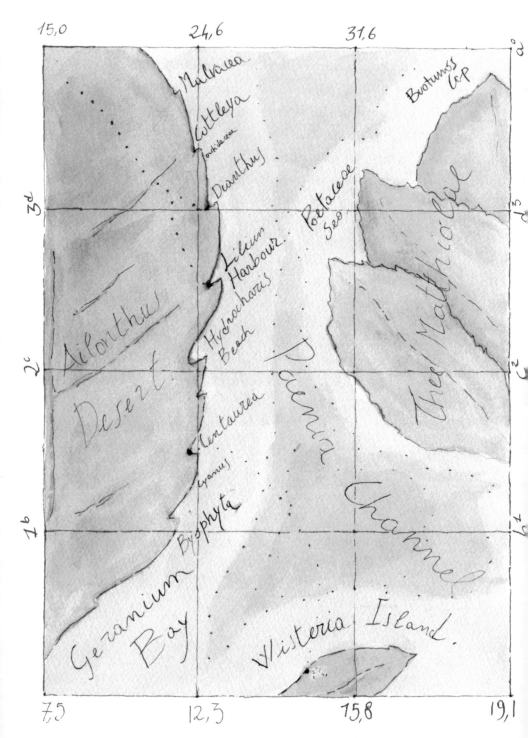

15,0   24,6   31,6   0ᵃ

Malvacea

Cattleya

Portulaca

Dianthus

Butomus Cap

Potaceae
Sea

3ᵈ   d⁵

Lilium
Harbour

The Matthiola Cat

Ailanthus

Hydrocharis
Beach

2ᶜ   c²

Desert

Paenia Channel

Centaurea

Cyanus

1ᵇ   b⁴

Bryophyta

Geranium

Bay

Wisteria Island.

7,5   12,3   15,8   19,1

6

ANACHRONISTIC,
LIKE AN ENCYCLOPEDIA

type species Avocado – domain Eukaryota – kingdom Plantae

division Magnoliophyta – class Magnoliopsida – subclass Magnoliidae

order Laurales – family Lauraceae – genus *Persea*

species *Persea americana* – origin Central America – diffusion Worldwide

first appearance in Europe Mid-sixteenth century

"Many are called but few are chosen." The Gospel
verse from Matthew describes perfectly the fate
of plant seeds. Enormous quantities of diaspores[62]
are produced annually, but only an insignificant
percentage of them survive. Some Lycopods ("wolf
feet," from the Greek *lykos*, "wolf," and *podos*,
"foot"), such as the *Lycopodium clavatum*, produce at least thirty
million spores per year, and it is nevertheless a rather rare species.
An Aleppo pine tree can produce from thirty to seventy thousand
seeds per year. Of these, probably fewer than two hundred will
make it to germination, and only a handful will survive. An
enormous production, whose final output practically approaches

zero. Such low percentages require strategies that improve, even if only a little, the seeds' chances of survival.

As we have seen, water, air, and animals are the carriers used by plants to disperse their seeds. The preference for one carrier over another is one of those evolutionary choices capable of influencing many characteristics of a plant, from its morphology to its physiology, from its capacity to adapt to its chances of ultimate survival. It is a choice that requires calculation and reflection. In making it, the characteristics of the carriers must be analyzed carefully. The so-called abiotic carriers, such as air and water, apart from small differences (I note this for the most persnickety of my readers), are the same in every place on Earth, and they maintain their characteristics pretty much unaltered over time. Water is always water, Captain Obvious might say, and wind can change in force and direction, but it does not transform or disappear over the centuries. In essence, air and water are two widely used carriers simply because they can always be counted on, in very different places, times, and circumstances. For this reason, even though their efficiency in seed dispersal is not that great and certainly inferior to the service offered by animals, air and water continue to be the preferred carriers of innumerable species. First of all, they are economical. They do not require the production of costly fruit in order to pay for animal services, and that is not something to be taken lightly. Plus, they are secure. This is a quality that counts a lot when it comes to turning over custody of your progeny. In all times and in every place on Earth, air and water are always ready to transport the seeds entrusted to them by plants.

It's a different story when you start to hand over your seeds to animals. The efficiency of the distribution is certainly better, while the level of security diminishes. How to decide whether to commit your savings to safe but low-yield investments or to others with high risk but much higher yields? Between these two extremes, there are many intermediate levels of risk and, consequently, profitability. In any event, it is an evaluation that calls for prudence. Some species choose security; others, profitability. Many decide, wisely, to diversify their investment, entrusting the dispersal of their seeds to two or more alternative systems.

Some plants, not wanting to run any of the risks implicit in entrusting their seeds to carriers, whatever they might be, have made a courageous decision that also distinguishes them from all of their vegetable colleagues. These plants, in fact, have decided to take direct responsibility for the entire process of dispersal, developing decidedly innovative and original instruments, such as, for example, explosive dispersal—a gimmick that is almost impossible to imagine in the peaceful and apparently inert world of plants. The species that entrust the fate of their progeny to an explosion are not very numerous, but those few that do literally make a lot of noise.

The Amazon forest and other tropical regions of Latin America are home to a powerful tree, the *Hura crepitans*, whose common name, the dynamite tree, leaves no doubt about its most fascinating feature. This species, which we can safely consider the queen of explosions, is able to project its seeds as far as 120 feet away at an initial velocity of more than two hundred feet per second.[63] These explosions can be so violent that the researchers tasked with gathering the seeds are forced to take cover behind

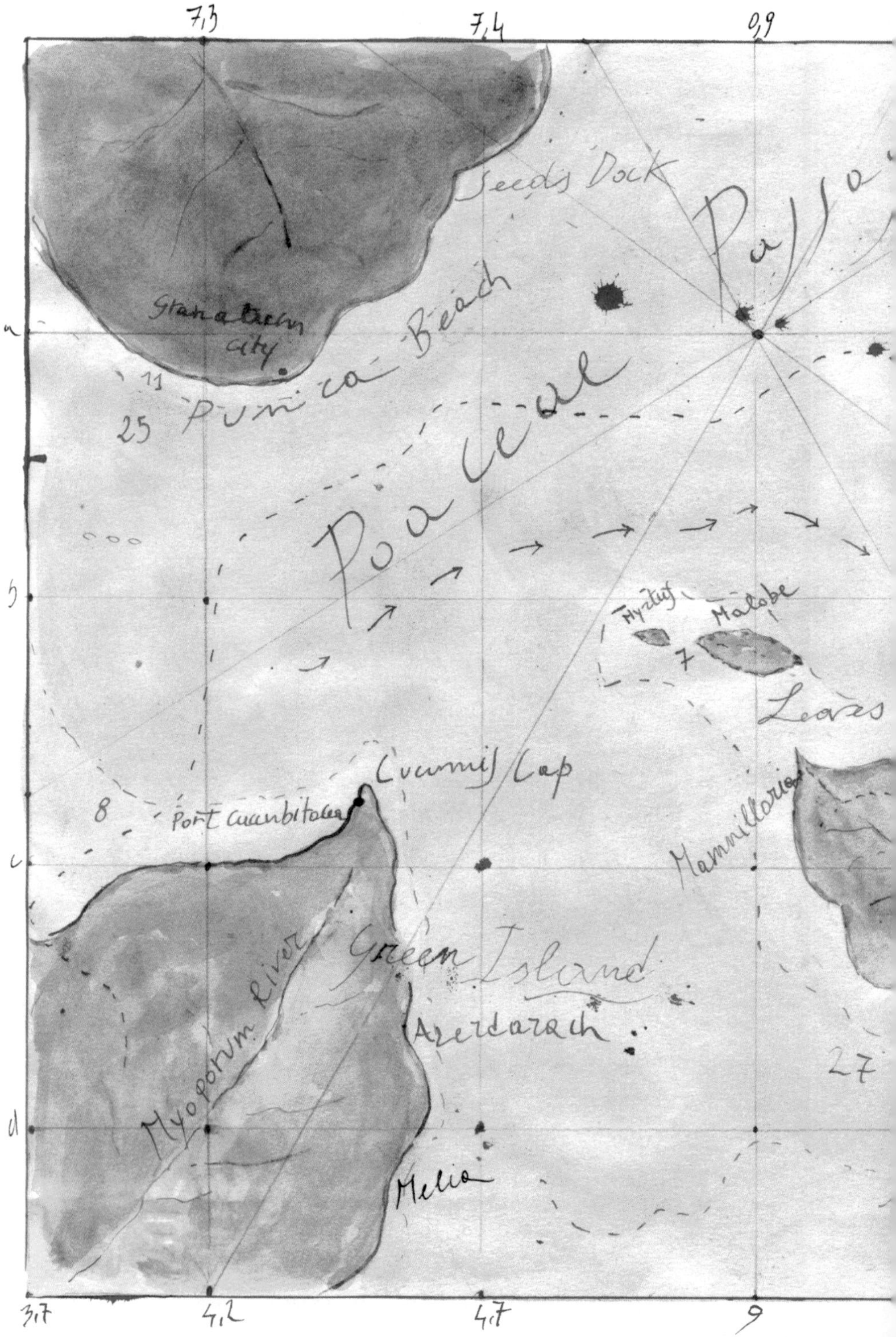
7,3
7,4
0,9
Seeds Dock
Pumica Beach
city
11
25
Poaceae
Malobe
7
Leaves
Cucumis Cap
Port Cucurbitacea
8
Mamillaria
Green Island
Myoporum River
Azedarach
27
Melia
3,7
4,2
4,7
9

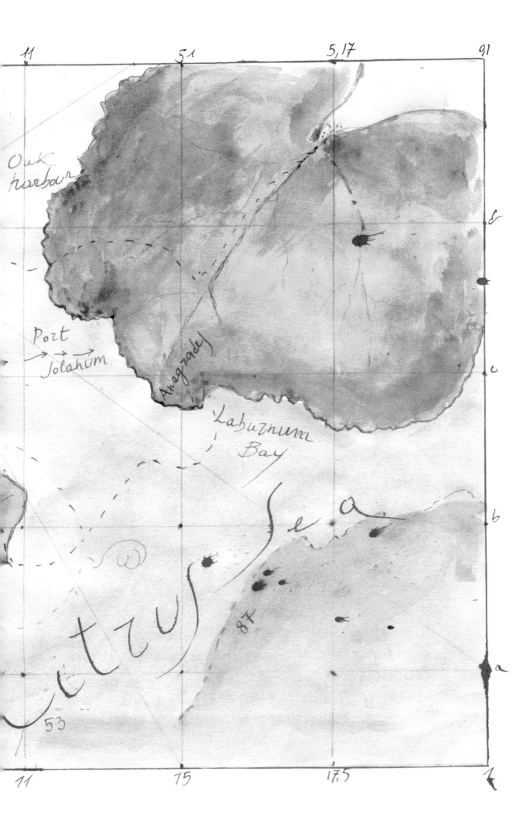

screens. Closer to home, many readers may have heard of the *Ecballium elaterium* (from the Greek *ecto*, "outside," and *ballo*, "throw"), the so-called squirting or exploding cucumber, able to project the mucilaginous liquid of its fruits, loaded with seeds, up to six feet away through a rapid explosive process. Another example is the quite common *Wisteria*, able to disperse its seeds by way of the sudden and powerful bursting of the pod that holds them. Explosive species are more numerous and widespread than is commonly believed.

Other species, instead of generating explosions, disperse their seeds by accompanying them delicately underground. The most famous of these seed-burying plants is the peanut (*Arachis hypogaea*), which inters its fruit during ripening, thus placing its seeds in the best possible conditions to germinate.

Even plants that choose animals as seed carriers find ways to attenuate their investment risk. Some species, for example, refrain from creating exclusive relationships, entrusting their seeds instead rather indiscriminately to any animal that happens to pass by. This is the case of the so-called hitchhiker seeds, which adopt various strategies and adaptations, such as claws, needles, and sticky surfaces, to attach themselves to animals in transit, which then transport them around the environment. In this case, the only requirement is that the animal has some fur that the seeds can grab on to. Plenty of animals possess this characteristic; there is no risk of remaining without a carrier. Other species rely on birds. In this case, too, the relationship is not exclusive. Any bird, of whatever species, as long as it's frugivorous, is welcome.

In other cases, however, plants construct special relationships with a limited number of animals. This kind of operation can be risky. Because of the high degree of specialization they require, these restricted relationships, though they ensure the best possible conditions for dispersal of a specific seed, can sometimes become precarious. While for seed dispersal we know of no certain cases of coevolution between plants and animals, such as those that do exist, for example, in the case of pollination, these relationships are nevertheless close ties between one vegetable species and a limited number of animal partners. If the animal, or group of animals, to which the plant entrusts the survival of its species happens for some reason to disappear, the plant also risks suffering the same fate.

This is exactly what happened to some plants that, having delegated the transportation of their seeds to specific animals that subsequently became extinct, found themselves at a certain point in their history faced with serious difficulties in the dispersal of their progeny. Some of these plants have disappeared just like their animal partners. Others have been saved by the skin of their teeth, while maintaining, however, as a souvenir of these "dangerous liaisons," some bizarre characteristics, reasonable only in light of their original partnership with now-extinct animals and today completely out of place. These vegetable adaptations to animals that no longer exist are called evolutionary anachronisms,[64] and they are much more common than one might think. Many species, for example, maintain adaptations made to defend themselves from or to attract now-extinct animals. Take holly (*Ilex aquifolium*), for example, a very common species. Having leaves with spindly edges up to a height of fourteen or fifteen feet is an anachronism. Once, when there existed in Europe huge herbivores able to nourish themselves on leaves born at remarkable heights, this defense certainly made sense, but in Europe today there are no animals able to nibble on leaves that high up.

As long as we're talking about anachronisms like the one just described, it's nothing serious. Such adaptations are pointless, but they do not impede the growth of the plant in any way. When, on the contrary, the anachronisms have to do with the delicate sphere of propagation, their consequences can be dramatic. Enormous seeds, for example, designed to be swallowed whole by animals that have not existed for millennia, are an anachronism capable of having a negative impact on a vegetable species' survival capacity. Some of these anachronistic species, having lost the animals that ensured their diffusion, have managed to survive all the same by weaving new relationships with different animals. A few have succeeded in entering into relationships with another very efficient and omnipresent transporter: humans. They have thus guaranteed themselves not only survival but also an unprecedented capacity for dispersal.

SURE DO MISS THOSE MASTODONS

Plants that produce large fruit, with lots of pulp, brightly colored, aromatic, and appetizing, do so for a reason. Investing so much energy in a casing whose only purpose should be to contain a seed, would seem to be totally inappropriate, if it weren't that these large fruits have quite different tasks to perform. They must serve as a lure, and at the same time as a reward, for all those animals that, by feeding on that fruit, perform the essential function of transporting the seeds far away from the mother plant. If a tree that produces large and attractive fruit lets it fall to rot at its feet, then something in its strategy for dispersal has gone awry. There can be no worse scenario for the survival hopes of a species.

Plants that do not entrust their seeds to animals normally develop minuscule fruit, often nearly invisible. If their seeds then have to be dispersed by wind, there is no use enlarging the size of their fruit. In fact, excessive size could turn out to be an obstacle to diffusion. On the other hand, plants that count on animals invest a lot of energy in fruit production. When a plant, despite this effort, cannot manage to disperse its seeds, it definitely finds itself in big trouble. The piling up of fruit at the foot of the mother plant means that the great majority of seeds will rot and lose their vitality. Even in the extremely lucky event that the seeds actually do manage to germinate, the plants they produce end up confronting a difficult environment, growing literally in the shadow of their mother, with very little available light and, therefore, scarce chances for survival. When fruit that falls from a tree is not eaten by an animal, it usually means that the animals for whom the fruit was intended no longer exist. Over time, in the absence of some unforeseeable ability to survive, these plants deprived of their animal partners are destined for extinction.

In nature everything is connected. This simple law that humans don't seem to understand has a corollary: the extinction of a species, besides being a calamity in and of itself, has unforeseeable consequences for the system to which the species belongs. Until about thirteen thousand years ago, for example, the Americas hosted a huge number of oversized animals. The quantity and variety of those animals is hard to imagine. If we could bring them back to life, as in the films by Steven Spielberg and others, their numbers would overwhelm us. Back in those days, there were oodles of giant sloths, various species of tapir, peccaries, giant camels like the *Titanotylopus* ten feet high at the shoulder, and then the *Bootherium* (musk-ox), the *Euceratherium* (shrub-ox), the elk moose, mammoths and mastodons without end, the *Glyptotherium*, giant beavers, *Hippidion*, quasi-armadillos such as the *Doedicurus* and the *Glyptodon*, behemoths such as the *Toxodon* or the *Stegomastodon*. And their predators, gigantic carnivores such as lions, *Smilodon*, *Homotherium*, and birds as big as Piper Cubs, like the *Teratornithidae*.

An entire world out of scale, inside of which we would have felt like Gulliver in Brobdingnag. Yet it seems that we little humans were actually the ones responsible for the abrupt extinction of this marvelous megafauna.[65] In the bat of an eyelash, these animals disappeared without a trace, except for the fossils, which have made it possible for us to tell their story. Some submit that it was climate change—even back then—but most scholars are agreed that the arrival of humans on the American continent was the proximate cause that led to the sudden extinction of animals that had trod the soil for tens of millions of years.

It is estimated that in North America alone, thirty-three genera of mammals describable as megafauna (animals with a body mass greater than ninety-seven pounds)[66] became extinct around thirteen thousand years ago. Through hunting, humans erased from the face of the Earth all of these giant-sized herbivores. Subsequently, the predators of these animals inevitably met the same fate, and, after an unstoppable chain of events, in the end there were none left. Plants could not remain immune from such a catastrophe.

When the discussion turns to extinction, the tendency is to speak only about animals. Plants are not considered, in part because we fail to understand their fundamental importance for life on the planet, and in part because, not leaving behind any bones, they are much more difficult to study. Determining the extinction of a vegetable species at a certain moment in history requires long and sophisticated analyses, usually based on minuscule grains of pollen.

Although plants are more adaptable than animals, many plants surely became extinct in concomitance with the disappearance of the megafauna. Many others suffered grave consequences from the animals' extinction but in the end managed to survive. Among the survivors are some well-known species, such as the persimmon and the papaya, and other lesser-known ones, like the *Maclura pomifera*. This plant, whose common name is the Osage orange—from the name of the Native American tribe that lived in the region where the tree grew—produces spherical infructescences of a diameter that can exceed six inches, much appreciated by the extinct North American herbivorous megafauna. Upon the disappearance of mastodons and mammoths, this species, too, inevitably found itself in trouble. For a certain period, some help for its diffusion came from wild horses, which started nourishing themselves on its fruit. In the end, with the continuous reduction as well of wild horses, salvation luckily arrived thanks to the plant's extremely hard wood, which made it the favorite of American cattle raisers for the construction of hedges and fences. If, rather than in 1874, barbed wire had been invented fifty years earlier, the Osage orange might no longer exist.

A species much better known than the Osage orange, which ran into the same problem and just barely managed to survive, is the avocado (*Persea americana*). Anyone who has cut open an avocado fruit cannot help but have noticed the enormous seed hosted in its center, like a Fabergé egg inside its luxurious case. An out-of-scale seed. Incomprehensible if we regard it as an instrument for the diffusion of its species. What animal, in fact, could ever swallow an entire avocado without damaging the seed inside it? Let's not forget, by the way, that swallowing a fruit is not enough to ensure the dispersal of the seeds of a vegetable species. It is essential that the seeds succeed in transiting undamaged through the animal's digestive tract. This requirement makes it so that many species, among them the avocado, defend their seeds by filling them with toxic substances that are released in the event they are damaged.

Today, American herbivores able to swallow an avocado fruit whole no longer exist, but up until thirteen thousand years ago, there were a slew of them—among them the *Gomphotherium*, an elephant species with four tusks; *Glyptodons*, giant armadillos ten feet long; and, finally, giant ground sloths such as the *Megatherium*, the size of today's elephants. All of them, by nourishing themselves on the fruit of avocados, facilitated the dispersal of their seeds. With their extinction, followed by that of all the other similar-sized herbivores, the avocado found itself, from one day to the next, without its principal partners and with an enormous seed that would have been hard to pass off to more modestly sized clients.

The fate of the species appeared to be sealed. Without its mastodons, the plant was destined to certain extinction. But one must never despair. You never know where help might come from. In the case of the avocado, it arrived from the most unexpected of animals: the jaguar. These carnivores, attracted by the oily pulp of the avocado, showed themselves, in fact, to be excellent carriers. Their teeth, made to tear meat rather than grind it, were perfect for avoiding damage to the precious seed of the avocado. Their jaws, accustomed to ingesting large chunks of meat, were well suited to swallowing whole avocados in a single bite. This couldn't be a definitive solution, but as a palliative, while the avocado waited to conclude a new contract with a more efficient disperser, jaguars were just fine. Thanks to them and a few other extemporaneous carriers, the avocado managed to stay alive, despite the inexorable shrinkage of its distribution area. And it would have kept on shrinking until vanishing altogether, if there hadn't appeared on the horizon, just as it seemed that all was lost, the perfect carrier: humans.

By the time the Spanish got to America, the avocado was limited to a very restricted area. Rescued in extremis, thanks to the appreciation of its fruit by the first European explorers, the species began to spread quickly all over the world. In 2016, the land area devoted to avocado cultivation amounted to more than 1,360,000 acres, spread over all the continents—an apparently unstoppable success. When the internet starts featuring articles entitled "Avocado Toast: Five Mistakes Not to Make" or, just to stay with five—a favorite number in web lists—"Five Ways to Eat Avocado Toast," it means that this fruit has entered the ranks of international cuisine. Indeed, year after year, the demand for avocados has been growing constantly, just like the amount of land devoted to its cultivation.

So then, everything's copacetic? Not by a long shot. Associating with humans is like making a pact with the devil. Sooner or later, you're asked to pay with your soul. That's what has happened with the avocado. And once again, it's the fault of that enormous seed that has been the cause of all its misadventures.

The same humans who, until a little while ago, successfully hunted those enormous saber-toothed tigers, have now become beings that find it unbearable to put up with the presence of seeds in fruit. They're such a bother. What are they doing in the middle of our food? So, as has already happened in the past to other species that have imprudently associated with humans, such as bananas, grapes, tomatoes, and citrus fruit, now it is the avocado's turn to satisfy a spoiled market by becoming seedless.

Once it has been deprived of the possibility to produce seeds, a plant is no longer a living being, but a simple means of production in the hands of the food industry, which decides how, how much, and where to reproduce it. That's not all. Without seeds, a plant cannot propagate through sexual reproduction but only vegetatively, producing daughter plants that are identical genetic clones of the mother plant. The genetic diversity of the species vanishes, and the same few individuals are propagated millions of times. A parasite or a disease that strikes one of these individuals is able to strike all of its clones. Just to take one example, nearly all of the bananas (without seeds, obviously) consumed in the U.S., Europe, and China come from the Cavendish cultivar.[67] Their genetic uniformity has made it so that a recently discovered fungal disease, to which the Cavendish is highly sensitive, now threatens much of the world's population of bananas.

We were talking about a pact with the devil? In 2017, a British supermarket chain began distributing packages of five seedless avocados called "cocktail avocados," which have the added advantage that they can be eaten without removing the skin. There you have it. Our children will not even imagine that once upon a time avocados had a seed inside, just as we have never seen a banana seed. Thus concludes, sadly, the parable of a great tropical tree: from the food of mastodons to cocktail party finger food. *Sic transit gloria mundi.*

THE DODO AND THE TAMBALACOQUE

The island of Mauritius is universally known as a sort of heaven on Earth. Today it is rather the worse for wear, but signs of its former beauty are conserved where the resorts have not arrived yet or in the less inhabited areas in the south of the island. Those who landed on Mauritius in the early years of the last century must have had the impression of arriving in an enchanted world. Between the seventeenth and nineteenth centuries, this island was a fundamental destination not only for botanists and naturalists but also for poets and writers, who broadcast its myth. Mark Twain wrote that "Mauritius and was made first, and then heaven; and . . . heaven was copied after Mauritius." Charles Baudelaire composed his first poem on this island, *À une dame créole*, while he was a guest at the botanical garden of Pamplemousses (the oldest

botanical garden in the tropics). Joseph Conrad, who knew the
island very well, having gone there frequently during his years as
a ship captain for the East India Company, described it as "a pearl
distilling much sweetness on the world."

Much of the island's charm, apart from its undeniable
natural beauty, which makes it the tropical island par excellence,
derives from its special history, populated by animals and
plants that have undergone here, undisturbed for millions of
years, their own parallel evolution. To visit Mauritius was to
visit an ongoing experiment in the possibilities of evolution. An
experiment that went on undisturbed until 1598, the year in
which the Dutch built their first settlement,[68] interrupting the
enchantment with the brutality typical of colonizers. When the
Europeans arrived on the island, Mauritius was populated by
the most fantastic flora and fauna that one can imagine. About
a third of the plants on the island were endemic, as were many
animal species. The island was a closed microcosm, with its
own rules, different from those of the places the colonists came
from. It was a world in which, for example, there not being
any large predatory animals, birds had evolved losing their

capacity to fly, becoming large, slow, and terrestrial. Peaceful and congenial birds, such as the legendary dodo—featured in *Alice in Wonderland*[69]—a magnificent columbiform bird, unable to fly, and weighing up to sixty-five pounds, were a numerous population on the island.

The descriptions of the first visitors to Mauritius recount a decidedly paradisiacal situation, with animals, headed up by the dodo, not at all afraid of this new biped guest, whose capacity for destruction would soon become evident. In the span of less than a century after the Dutch settlement, the entire dodo population of Mauritius—and, therefore, of the world—had been wiped out,[70] partly by way of pointless hunting (it seems that the dodo's meat was disgusting), partly through the elimination of its habitat in favor of enormous sugarcane plantations. Finally, because it was the prey of animals like pigs and dogs introduced to the island's delicate ecosystem by humans. The same sad fate was suffered by the broad-billed parrot and dozens of other species, including the magnificent gigantic sea turtles of Mauritius, of which the only remaining testimony are some enormous shells and a certain number of prints showing them being ridden by Dutch soldiers, sitting comfortably atop their shells.

On the island of Mauritius, as I have said, the rules were different from those in the rest of the world, dictated by an evolution that had followed its own original paths. Paths that had led to a situation in which the main pollinator of flowers on the island was a blue gecko, while seeds were dispersed by gigantic sea turtles, broad-billed parrots, diurnal bats, and, obviously, by the dodo. With their sudden extinction, many plants found themselves deprived of their principal partners for the dispersal of their seeds. These plants included a tree endemic to the island, called by the French tambalacoque (*Sideroxylon grandiflorum*, known until a few years ago as *Calvaria minor*).

In 1977, an American ornithologist, Stanley Temple, provoked a heated discussion in the scientific community following the publication of an article of his in *Science*, in which he suggested that there was a mutualistic relationship between the tree and the dodo.[71] Temple claimed that in order for the seeds of the tambalacoque to germinate, they had first to go through the digestive system of the dodo. There, the combined abrasive action of the bird's gizzard and its stomach acids, by corroding the wood surface of the seeds, allowed water to penetrate them, setting off the germination process. It followed that, with the extinction of the dodo bird, the tree, too, would necessarily have been destined to extinction. In support of his argument, Temple produced two rather solid pieces of evidence. The first concerned the number of trees present on the island. In 1977, according to Temple, there were only thirteen trees left, and each of them, in his judgment, was more than three hundred years old. Therefore, they were the last trees germinated before the definitive extinction of the dodo toward the end of the seventeenth century. The second piece of evidence was of a more experimental nature. Temple had identified turkeys as a bird endowed with a gizzard that was, in certain ways, similar to the dodo's. Based on this similarity, he decided to have turkeys ingest seventeen tambalacoque seeds, and, after recovering them from the turkeys' feces, he succeeded in getting three of them to germinate.

This theory had an indisputable charm about it and seemed

reasonable. Moreover, the fact that it had been published by an authoritative journal like *Science* helped spread the theory quickly all over the world. In the years that followed, a series of studies demonstrated the partial invalidity of the theory. First of all, upon a more in-depth analysis of the vegetation, it turned out that there continued to exist on the island many more than the thirteen trees counted by Temple and, above all, that many of them were much younger than the three hundred years necessary to support his theory. This is not to say that the tree was flourishing. It is a species on the road to extinction, whose number of surviving members, albeit far greater than the thirteen found by Temple, is nonetheless well under the minimum number necessary to ensure the species a decent chance of survival. The extinction of the dodo, like that of many other frugivorous animals involved in the dispersal of seeds, was surely influential. The destruction of the tree's original habitat, in large part replaced by plantations of sugarcane and coconut palms, did the rest.

The survival of species is a very delicate thing, and environmental changes tied to human activity have shown themselves in the past—and will show themselves even more so in the coming years—to be deleterious for a vast number of living organisms. Especially animals, less adaptable than plants, according to what the latest studies on the subject appear to demonstrate.[72]

Even if the story of the ineluctable bond between the dodo and the tambalacoque turned out in the end not to be altogether true, Temple's work had the merit of revealing the lack of knowledge and, in general, of interest about the effects that animal extinction could have on vegetable life. Following publication of Temple's article, a growing number of researchers started working on the issue and unique relationships between plants and animals began to emerge and to be studied in the depth and detail they deserved.

One animal that entertains a large number of unique relationships with vegetable species is the elephant. Many seeds of African flora seem to depend on transiting through the digestive system of these pachyderms in order to be able to germinate. The *Omphalocarpum elatum* (*Omphalocarpum*, "umbilical fruit": take a look at the fruit to understand the name), for example, is a relative of the tambalacoque, both being members of the Sapotaceae family, and even shares with it a special predisposition to creating stable ties with animals. It is a species original to central Africa, and it is utterly unmistakable. The tree, in fact, produces voluminous and heavy fruit, weighing about five pounds, directly on its trunk.

This is not, however, the most unusual characteristic of these plants. Their fruit is practically indestructible, protected by an armor that no animal except an elephant is able to split open. The technique used by the elephants has been documented only recently. It calls for the fruit to be speared by a tusk and then split in two by pressing it between the ground and the base of the tree, a complex procedure that no other animal has the chance or capacity to perform. In this case, the relationship between the tree and the elephant is such that the resonating sound of the fruit falling on the ground is enough to call the elephants, who, through special passages carved out of the compact mass of the forest, come running to the banquet. Were the elephant to become extinct, the *Omphalocarpum elatum*, like so many other vegetable species that depend on its diffusion, would suffer the same fate. Every living species is part of a network of relationships about which we know very little. Therefore, every living organism must be protected. Life is a rare commodity in the universe.

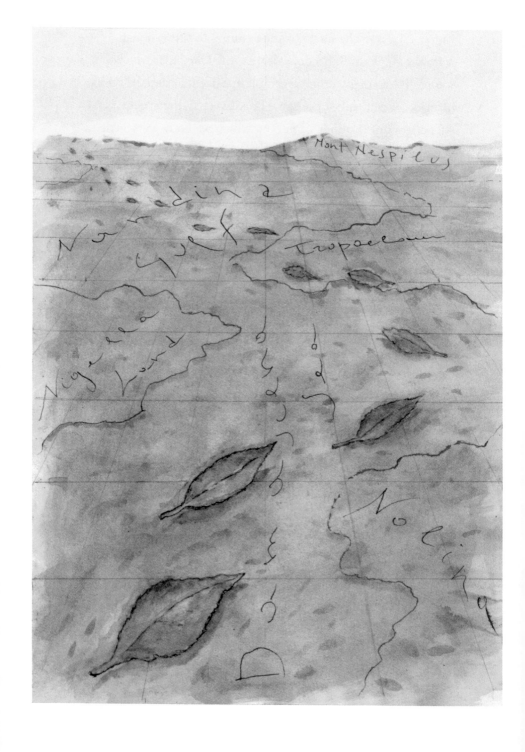

· NOTES ·

1. Pioneers, Combatants, and Veterans

1. The name *Asplenium* comes from the Greek *splen* (spleen). In ancient times these ferns were used as a remedy for diseases of the spleen. The other meaning of "spleen," the "existential angst" tied to the sensitive nature of the poet, made famous by Charles Baudelaire, also derives from *splen*. According to Hippocrates' theory of the humors, the black bile produced by the spleen could give rise to a state of disquiet, unease, and ennui. This seems to have led to the belief that the little *Asplenium* ferns are useful for treating spleen.

2. Robert Decker and Barbara Decker, *Volcanoes* (New York: W. H. Freeman, 1997).

3. Þórarinsson Sigurður, "The Surtsey Eruption: Course of Events and the Development of the New Island," in *Surtsey Research Progress Report* I, (Reykjavik: The Surtsey Research Society, 1965): 51–55.

4. One famous case concerns the island of Ferdinandea, which emerged off the coast of Sicily in 1831, after an underwater

eruption. The island grew to a surface area of 1.5 square miles and a height of 213 feet, but it didn't live long. Composed of an eruptive rock material easily eroded by wave action, Ferdinandea vanished underwater in January 1832.

5. Halophytes are rather rare; fewer than 2 percent of plants have this capacity.

6. Jonathan D. Sauer, *Plant Migration. The Dynamics of Geographic Patterning in Seed Plant Species* (Berkeley: University of California Press, 1988).

7. Thomas D. Brock, "Primary Colonization of Surtsey, with Special Reference to the Blue-Green Algae," *Oikos* 24 (1973): 239–43.

8. Sturla Fridriksson and Haraldur Sigurdsson, "Dispersal of Seed by Snow Buntings to Surtsey in 1967," in *Surtsey Research Progress Report* IV (Reykjavik: The Surtsey Research Society, 1968): 43–49.

9. Sturla Fridriksson, "Plant Colonization of a Volcanic Island, Surtsey, Iceland," in *Artic and Alpine Research* 19, no. 4 (1987): 425–431.

10. This is the message that all the inhabitants of Pripyat heard on the day of the evacuation: "Attention, attention! Attention, attention! Attention, attention! Attention, attention! The City Council hereby informs you that, following the accident at the Chernobyl nuclear power plant, the conditions of the atmosphere surrounding the city of Pripyat are proving to be harmful with high levels of radioactivity.
    "The Communist Party, its officers, and the armed forces are therefore taking all the necessary measures. Nevertheless, with the aim of ensuring the total safety of the people, and above all children, it has become necessary to evacuate the citizens temporarily to nearby cities and towns in the region of Kiev. For this purpose, today, April 27, starting at 2:00 p.m., buses will be sent under the supervision of the police and city officials. You are urged to bring with you your papers, strictly necessary personal effects, and essential food items.
    "The top management of the city's public institutions and

industry have drawn up a list of employees required to remain in Pripyat to ensure the normal functioning of the city's businesses. Furthermore, all habitations, during the period of evacuation, will be under police surveillance.

Comrades, as you are leaving your homes temporarily, please do not forget to close the windows, turn off all electric and gas appliances, and shut off the water. Please maintain calm, order, and discipline during the execution of this temporary evacuation."

11. Katerina Klubiková et al., "Proteomics Analysis of Flax Grown in Chernobyl Area Suggests Limited Effect of Contaminated Environment on Seed Proteome," *Environmental Science & Technology* 44, no. 18 (2010): 6940–46.

12. Dharmendra Kumar Gupta and Clemens Walther, eds., *Radionuclide Contamination and Remediation Through Plants* (Berlin-Heidelberg: Springer, 2014).

13. The LINV is an international laboratory that I founded in 2005. Those interested in more detailed information can visit the website www.linv.org.

14. The compound noun *Hibakujumoku* is composed from *hibaku*, "bombarded, exposed to nuclear radiation," and *jumoku*, "tree" or "forest."

15. A *bento* is a container, traditionally made of wood, with a lid, used to carry a complete meal, to be eaten at home or in the open air. It is an object of everyday use in Japan.

2. Fugitives and Conquerors

16. Bruce F. Benz, "Archaeological Evidence of Teosinte Domestication from Guilà Naquitz, Oaxaca," *PNAS* 98 (2001): 2104–106.

17. In *The English Physician Enlarged: . . . Being an Astrologo-Physical Discourse of the Vulgar Herbs of This Nation . . .* , 1652.

18. Cynthia S. Kolar and David M. Lodge, "Progress in Invasion Biology: Predicting Invaders," in *Trends in Ecology and Evolution* 16 (2001): 199–204.

19. The correct term for this capability is "phenotypic plasticity."

20. According to The Plant List (http://www.theplantlist.org), a site created and maintained by two of the principal botanical institutions in the world: the Royal Botanic Gardens, Kew, and the Missouri Botanical Garden.

21. The pappus is a feathery appendix characteristic of some fruits and seeds, whose main function is to favor the dispersal of seeds by wind.

22. Stephen A. Harris, "Introduction of Oxford Ragwort, 'Senecio squalidus L. (Asteraceae),' to the United Kingdom," in *Watsonia* 24 (2002): 31–43.

23. In 1795, when the botanical garden of Palermo was founded, more than two thousand plants were moved there from the Misilmeri garden.

24. One variety of the *Pennisetum setaceum*, called *rubrum*, has won the most coveted prize for ornamental plants: the Award of Garden Merit, awarded since 1922 in the United Kingdom. Before the awards are assigned, the competing plants are made to grow for one or two years in the climatic conditions of Great Britain. The trial reports are made available as brochures on the website. The awards are reexamined annually in case the plants have for some reason become unavailable on the market or been replaced by better cultivars.

25. Salvatore Pasta, Emilio Badalamenti, and Tommaso La Mantia, "Tempi e modi di un invasione incontrastata: 'Pennisetum setaceum (Forssk.) Chiov. (Poaceae)' in Sicilia," *Naturalist siciliano* IV, XXXIV, 3–4 (2010): 487–525.

26. ISSG database, http://www.invasivespecies.net.

27. Pia Parolin, Stephanie Bartel, and Barbara Rudolph, "The Beautiful Water Hyacinth *Eichhornia crassipes* and the Role of Botanic Gardens in the Spread of an Aggressive Invader," *Boll. Mus. Ist. Biol. Univ. Genova* 72 (2010): 55–66.

28. Legendary spy (known by the name of "black panther"). During the Boer War he was given orders to assassinate Burnham and in 1942 was incarcerated as a spy for the Third Reich.

29. Jon Mooallem, "American Hippopotamus," *The Atavist Magazine*, no. 32, December 2013, https://magazine.atavist.com/american-hippotamus.

3. Captains Courageous

30. Charles Darwin, "On the Action of Sea-Water on the Germanization of Seeds," *Botanical Journal of the Linnean Society* 1 (1856): 130–40.

31. One Indian sect, for example, had as a religious precept the duty to plant coconut palms on every newly discovered atoll or island in the Pacific so they could ensure the survival of voyagers who happened to stop there. The story is cited by Emilio Chiavenda in his fundamental contribution to our knowledge of coconut palms, published in two parts: "La culla del cocco," *Webba* 1 (1921): 199–294, and *Webba* 2 (1923): 359–449.

32. Tribe of plants of the Arecaceae family, whose member species include the coconut palm tree (*Cocos nucifera*).

33. Jonathan S. Freidlander et al., "The Genetic Structure of Pacific Islanders," *PLoS Genetics* 4, no. 1 (2008).

34. Pablo Muñoz-Rodriguez et al., "Reconciling Conflicting Plylogenies in the Origin of Sweet Potato and Dispersal to Polynesia," *Current Biology* 28 (2018): 1246–56.

35. *Lodoicea maldavica* grows only in the Seychelles. And, therefore, it is endemic (or exclusive) to that archipelago.

36. Robert Nieuwenhuys, *Mirror of the Indies: A History of Dutch Colonial Literature* (Amherst: University of Massachusetts Press, 1982).

37. Bianca A. Santini and Carlos Matroell, "Does Retained-Seed Priming Drive the Evolution of Serotiny in Drylands? An Assessment Using the Cactus 'Mammillaria Hernandezii,'" *American Journal of Botany* 100, no. 2 (2013): 365–73.

38. Suzanne Simard et al., "Resource Transfer Between Plants Through Ectomycorrhizal Fungal Networks," in Thomas R. Horton, ed., *Mycorrhizal Networks* (Dodrecht: Springer, 2015), 133–76.

39. Peter J. Edwards et al., "The Nutrient Economy of *Lodoicea maldivica*, a Mondominat Palm Producing the World's Largest Seed," *New Phytologist* 206, no. 3 (2015): 990–99.

4. Time Travelers

40. Ulf Büntgen and Nicola di Cosmo, "Climatic and Environmental Aspects of the Mongol Withdrawal from Hungary in 1242 CE," in *Scientific Reports* 6, 26 May 2016, www.nature.com/articles/srep25606.

41. It was the first, and the longest lived, among the sister republics of the French Republic. In fact, it was a satellite state of France.

42. The name comes from the short, curved sword, of Thracian origin, also used by the Romans and called a *sica*.

43. Flavius Josephus writes in *The Jewish War*: "They call it the snake because of its narrowness and constant windings: it is

broken as it rounds the projecting cliffs and often turns back on itself, then, lengthening out again a little at a time, manages to make some trifling advance. Walking along it is like balancing on a tightrope. The least slip means death; for on either side yawns an abyss so terrifying that it could make the boldest tremble. After an agonizing march of three and a half miles the summit is reached, which does not narrow to a sharp point but is a sort of elevated plateau." Translation by G. A. Williamson.

44. Gwynn Davies, "Under Siege: The Roman Field Works at Masada," *Bulletin of the American Schools of Oriental Research* 362 (2011): 65–83.

45. Arie S. Issar, *Climate Changes During the Holocene and Their Impact on Hydrological Systems* (Cambridge: Cambridge University Press, 2003).

46. Sarah Sallon et al., "Germination, Genetics, and Growth of an Ancient Date Seed," *Science* 320 (2008): 1464.

47. Jane Shen-Miller et al., "Long-Living Lotus: Germination and Soil Gamma-Irradiation of Centuries-Old Fruits and Cultivation, Growth, and Phenotypic Abnormalities of Offspring," *American Journal of Botany* 89 (2002): 236–47.

48. Norman Polmar, "Stalin's Slave Ships: Kolyma, the Gulag Fleet, and the Role of the West" (review), in *Journal of Cold War Studies* 9 (2007): 180–182.

49. Svetlana G. Yashina et al., "Viability of Higher Plant Seeds of Late Pleistocene Age from Permafrost Deposits as Determined by 'in vitro' Culturing," *Doklady Biological Sciences* 383 (2002): 151–54.

50. The land bridge across the Bering Strait, also called Beringia, was an isthmus at most a thousand miles wide that connected Alaska to Siberia during the Pleistocene Ice Ages.

5. Solitary Trees

51. Luis W. Alvarez et al., "Extraterrestrial Cause for the Cretaceous-Tertiary Extinction," *Science* 208 (1980): 1095–108. Luis W. Alvarez was also one of the scientists who flew on the bomber *The Great Artiste* on August 6, 1945, to observe the effects of the explosion of the atomic bomb on Hiroshima.

52. Paul J. Crutzen, "Geology of Mankind," *Nature* 415 (2002): 23.

53. Will Steffen et al., "The Trajectory of the Anthropocene: The Great Acceleration," *The Anthropocene Review* 2, no. 1 (2015): 81–98.

54. Gerardo Ceballos, Paul R. Erlich, and Rodolfo Dirzo, "Biological Annihilation Via the Ongoing Sixth Mass Extinction Signaled by Vertebrate Population Losses and Declines," *PNAS* 114 (2017), https://doi.org/10.1073/pnas.1704949114.

55. Simon Lewis and Mark A. Maslin, "Defining the Anthropocene," *Nature* 519 (2015): 171–80.

56. Paul J. Crutzen and Eugene F. Stoermer, "The 'Anthropocene,'" *Global Change Newsletter*, 41 (2000): 17–18.

57. Chris S. M. Turney et al., "Global Peak in Atmospheric Radiocarbon Provides a Potential Definition of the Onset of the Anthropocene Epoch in 1965," *Scientific Reports* 8 (2018).

58. From the Arabic *mahrī*, which refers to a very fast dromedary.

59. Local legends locate the earthly paradise precisely in the archipelago of Bahrain. This claim is not all that rare. There are countless countries around the world that claim to be the location of the original Eden. What those other countries do not have is a tree of life that is still alive and, all things considered, in excellent condition.

60. In 1960, near Tucson, Arizona, roots of a *Prosopis juliflora* were discovered at a depth of 174 feet. See Walter S. Phillips, "Depth of Roots in Soil," *Ecology* 44, no. 2 (1963): 424–67.

61. In the 1950s the *Prosopis juliflora* and other species of the same genus were again imported to Bahrain and used, this time with much greater success, for extensive reforestation.

6. Anachronistic, Like an Encyclopedia

62. Any part of a vegetable capable of ensuring its multiplication: spores, seeds, fruits, pieces of fruit, propagules, etc.

63. Mike D. Swaine and Thomas Beer, "Explosive Seed Dispersal in 'Hura Crepitans L. (Euphorbiaceae),'" *New Phytologist* 78 (1977): 695–708.

64. "Evolutionary anachronism" is a concept in evolutionary biology whose general theory was first formulated by the botanist Daniel Janzen and the geologist Paul S. Martin in an article entitled "Neotropical Anachronisms: The Fruits the Gomphotheres Ate," published in *Science* in 1982.

65. Mark A. Carrasco et al., "Graham Qualifying the Extent of North American Mammal Extinction Relative to the Pre-Anthropogenic Baseline," *Plos One* 4, no. 12 (2009), https://doi.org/10.1371/journal.pone.0008331.

66. Paul. S. Martin and Richard Klein, eds., *Quarternary Extinctions: A Prehistoric Revolution* (Tucson: University of Arizona Press, 1984).

67. United Nations Food and Agriculture Organization, www.fao.org/economic/est/est-commodities/bananas/bananafacts/en/#.Xdv-a5NKg3i

68. The island was already known to the Arabs, who called it *Dina Arabi*, as early as the tenth century. The Portuguese landed there in 1505, calling it *Ilha do Cerne* (Swan Island), but in fact it remained uninhabited until the first Dutch settlement in 1598.

69. "Lewis Carroll" was the pseudonym of Charles Lutwidge Dodgson. With the dodo in *Alice*, Carroll represents himself in the act of pronouncing his real name, "Do-Do-Dodgson." It is known that the author suffered from a mild form of stuttering.

70. Anthony Cheke and Julian Hume in *Lost Land of the Dodo: The Ecological History of Mauritius, Reunion, and Rodriguez* (New Haven: Yale University Press, 2008) identify 1662 as the year the last dodo died. Other sources indicate 1681.

71. Stanley A. Temple, "Plant-Animal Mutualism: Coevolution with Dodo Leads to Near Extinction of Plant," *Science* 197 (1977): 885–86.

72. Matthias Schleuning et al., "Ecological Networks Are More Sensitive to Plant Than to Animal Extinction Under Climate Change," *Nature Communications* 7 (2016).